David N. Griffiths

Implementing Quality

With a Customer Focus

Implementing Quality

With a Customer Focus

David N. Griffiths

Quality Press
American Society for Quality Control
Milwaukee, Wisconsin

Quality Resources
A Division of The Kraus Organization Limited
White Plains, New York

Printed in the United States of America

94 93 92 91   10 9 8 7 6 5 4 3 2 1

ASQC Quality Press
310 West Wisconsin Avenue
Milwaukee, Wisconsin 53203

Quality Resources
A Division of The Kraus Organization Limited
One Water Street, White Plains, New York 10601

Library of Congress Cataloging-in-Publication Data
Griffiths, David, 1935–
    Implementing quality with a customer focus / David Griffiths.
    p. cm.
    Includes bibliographical references and index.
    ISBN 0-527-91648-X
    1. Gas industry—Management—Case studies. 2. Gas industry—
Quality control—Case studies. 3. Gas industry—Customer services—
Case studies. I. Title.
HD9581.A2G75 1991
363.6'3'0688—dc20   90-20987
                                                        CIP

To Barbette, Michael, and Megan
for their love and support.

Contents

Acknowledgments

There are many to thank for their contributions to these writings. In particular, I want to recognize Margot Finot who "came out of retirement" to once again assist me; Don Lindemann, CEO of Citizens Gas & Coke, who leads by example and "walks the talk"; Joe Gufreda, senior manager of Ernst and Young, who for three years has been my facilitator, coach, and teacher; and most importantly, the men and women of Citizens Gas & Coke Utility who are proving daily that quality can be implemented with a customer focus.

Special recognition is also extended to Xerox Corporation, and in particular, to Mr. John Kelsch, Xerox Quality Director, who so readily shared their "Leadership Through Quality" approaches with us. The customer satisfaction process methodology described in this book is from their teachings. It is with sincere appreciation that I acknowledge their help and salute them as 1989 winners of the Malcolm Baldrige Award—and as our quality mentors.

Introduction

Citizens Gas & Coke Utility is a gas-distribution company supplying 225,000 customers in Marion County (Indianapolis), Indiana. We also manufacture gas that is mixed with purchased natural gas to form our main product. Significant byproducts of our manufacturing operation are foundry- and blast-furnace cokes. We are both a service company and a manufacturing company. One side of our business is regulated, the other is not. We employ approximately twelve hundred people, 750 of whom are in our gas division.

Four years ago we began our quality pursuit for reasons described by Don Lindemann, our chief executive officer, as being the "Three Cs": Customers (to be more acutely aware of them), Competition (to be more responsive to competitive challenges), and Change (to be better able to anticipate and respond). To this early rationale I now add two more "Cs" to the why of quality implementation: Competence (to tap the underutilized resource of modern-day employees) and Common sense (to recognize the most logical way to operate any organization).

Our focus begins with the customer, both external and

internal, whom we define as "anyone using our products, services, or outputs." We define quality as "satisfying the needs of our customers."

This book was written to share our experiences and the techniques we used as we developed and implemented our company-wide quality effort. It is intended as a step-by-step guide and commentary about the philosophy and attitudes I think necessary for successfully implementing quality. It describes not just what to do, but *how* to do it.

The attitudinal part of the equation may be the most important. Without it, in proper form and substance, I question whether an organization will have the durability to maintain the effort.

A certain feeling must be present within those who lead the transformation, and must be conveyed to all employees to the point of actually being instilled within them. This feeling for and about quality provides the atmosphere in which the tools and techniques make sense.

A company's commitment must embrace quality not in the form of a program or project, but as a *process*—ever dynamic, always changing—with a beginning but no end. Thus the more you do, the more there is to be done. Yet, in this truism is the opportunity for continual improvement.

Implementing quality is not without struggle or frustration. We have been taught to expect or to cause absolute results within specified time periods. The pursuit of total quality requires faithful concentration upon the process and its delivery mechanisms, not at the expense of results, but with a trust that recognizable achievement will be a natural outcome of this concentration. This shift in emphasis is difficult for many to accept, yet the abundance of what in fact flows as a result of such a shift is without boundaries.

Hopefully, the reader will find a helpful mix of "what" to do with a large quantity of "how to do it."

Part I

Investigating, Selecting,
and Developing
Your Quality Concepts

Chapter 1

Establish a Customer Focus

EVERYONE MUST FOCUS ON CUSTOMERS

Most organizations have elements of the quality process already in place or, at the very least, have tried programs or projects embracing quality characteristics. Efforts targeting problem solving, productivity improvement, employee involvement, and/or quality assurance are not uncommon corporate objectives. However, a single recognizable common goal can provide the catalyst for all employees to become earnestly involved in the pursuit of total quality. Too often our organizational pursuits are lacking in a recognizable and, more importantly, understandable focus. The truth is we aren't very good at presenting *why* work tasks are done. Quality can best be defined as satisfying customer needs, both externally and internally. Without exception, I believe "to satisfy customers" frames all work and enables each of us to understand why we're doing our task. This is as applicable to a manufacturing concern as it is to a service enterprise.

B. Joseph White, Ph.D., professor of business administration, University of Michigan, sums it up when he states: "First, focus on the customer. The purpose of all work and of all improvement efforts is to better serve customers."

Focusing on anything other than the customer has the potential for excluding many, if not most, employees. Productivity improvement may be seen as a management ploy to the disadvantage of the rank and file; quality improvement and its use of statistical techniques may be meaningful only to a handful. Employee involvement may be seen as warm and fuzzy (and thus without substance), which is probably why so few organizations implement long-enduring, organization-wide quality.

QUALITY IS A PROCESS

Quality must be viewed as a process, not a program or a project. It has a beginning but no end. It encompasses all, is all-consuming. Quality has the power to energize an organization, yet its concepts are so simple that most will exclaim "I know that!" when introduced to it.

Why is it then so few master the process, and those who do stand out above all others? What causes some organizations to achieve so much more than the ordinary? I believe that only those who totally embrace the customer and work to satisfy the customer's needs will become quality masters. Only when satisfying customer needs is made the focus, can an organization view all its activities within their proper perspective.

There was a time when the development of corporate and departmental action plans was almost exclusively budget-driven. As management techniques became more sophisticated, greater emphasis was placed upon long-range target setting (strategic planning) to develop goals and objectives for which strategies, action steps, and budgets were formulated. A further maturing brought recognition that organizations and their endeavors were better served if there was a stated philosophy or mission providing an operational cornerstone and, better yet, a feeling about what the organization was doing. Even though the words "customer" and "customer satisfaction" are routinely found in these mission statements, usually the actual analysis of customer needs and expectations is left undone.

CUSTOMER-NEEDS ASSESSMENT

Adding customer-needs assessment and recognizing customer perceptions and expectations to the management process (Figure 1.1) provides a dynamic power to developing and implementing action plans. When we ask "Who is the customer?" and "What are the customers' needs and/or expectations?" we identify "what" we should be doing. The organization's mission, targets, and strategies can then be considered in terms of satisfying the customer, and the entire management process is brought together to work toward a common goal.

The customer-needs assessment should never be considered only in the organizational vacuum of our own experience or knowledge of customer needs. The inclination is to think we know what the customer needs, but there is only one sure way to assess customer needs and that is to *ask* the customer to provide the information. Communication between the customer and the organization is essential, and a dialogue must be initiated to provide regular (not just an annual survey) identification of needs and measurement of accomplishment and satisfaction.

You may recall I define quality as satisfying customer needs not only externally but internally as well. There is great worth in external customer assessment, but the concept of applying the same considerations to internal customer assessment is at least equal to, and perhaps even exceeds, the potential of external examination.

Analyzing what I do (my outputs), who I do it for (my

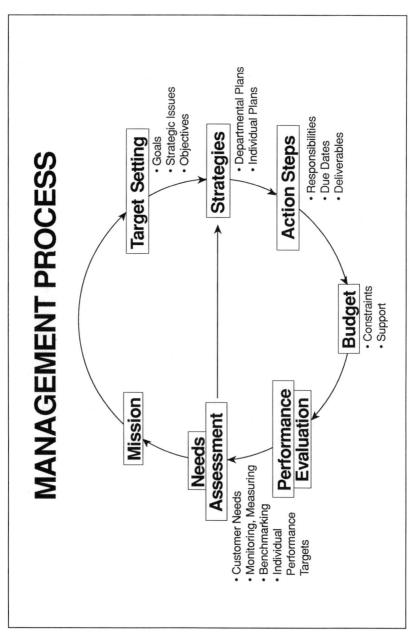

FIGURE 1.1 Recognizing customer perceptions and expectations is part of the management process.

customers); and what is expected (customer needs) by every employee in a company brings new experiences and meaningful change to the workplace. Fiefdoms don't exist comfortably in an internally customer-focused organization. When people begin to try to understand their jobs in the context of satisfying customer needs rather than in a robotic adherence to job-description tasks, performance improves.

It is all so simple. Yet, too often we're "too busy" and "don't have the time" to communicate with each other . . . to really find out what's needed or how what we provide is being perceived. We tell ourselves we already know . . . either because we haven't heard differently or because it's the way we've always done it. However, if we reach out, we may find some of our outputs aren't at all what they should be or could be. In some instances we may find we're doing some things that aren't even required or (heaven forbid) for which there is no customer.

White states it all nicely: "Quality (remember, quality is satisfying customer needs internally as well as externally) is a goal people rally around, unlike other operational goals like cost reduction or productivity improvement. Quality opens people up to change because the change is for a good reason. It connects them with the customer and taps the motive of pride in their work."

Establish a customer focus in the beginning of your quality effort, and you will have provided a solid foundation upon which to build.

Chapter 2

Commit to the Quality Process

LEADERSHIP THROUGH QUALITY

Without top management commitment to the quality process, there is little likelihood it will survive. David Kearns, former chairman and CEO of Xerox, had this to say as Xerox embarked upon its "Leadership through Quality" journey in the early 1980s: "We're going to do it . . . whatever 'it' is." That's the type of commitment it takes. Ideally, the commitment should be from the CEO, but quality can be embraced by any department or work unit. Even if quality concepts are applied by the individuals, they will soon be viewed as head and shoulders above others in performance. Also, quality can be practiced everywhere, not just in the workplace.

Faith and Discipline

Quality doesn't just happen. It must be made to happen. It requires faith and discipline. Faith that goes beyond simple bottom-line justification. Faith in the belief that if you adhere to the process, positive benefits will result. (They do, too. Citizens Gas & Coke had one team that, in the first year of the quality-improvement process, produced annual savings exceeding all out-of-pocket costs of the first three years.) Faith that acknowledges that the quality way is the right way to do business. Faith, admittedly, is the "soft" side of quality. It is an intangible, but it is the "feeling" about quality that will carry you through those days when progress is leading you temporarily backward rather than forward.

Discipline is another ingredient of commitment. Having now experienced four years of the quality pursuit, I'm struck with how undisciplined we are in most management practices. Some may argue this point, but I believe we too often hear the wrong voices (usually our own), don't follow a well-defined process in providing our outputs, generate solutions without analyzing the issues, and measure the wrong things. It takes discipline to keep the organizational house in order, to formulate and use the appropriate process, to involve others, to listen, and to be a leader, not a boss.

Commitment also means striving to be the best—continually trying to improve. It has to it a thirst for information about who might be doing something better than you and a desire to know how it is done. A commitment to quality improvement instills a willingness to share information and experiences.

MAKING QUALITY PERSONAL

Committing to the quality process is done by making it more than a policy or procedure. It is accomplished by application and by deed. You cannot invoke quality from on high, handing down its tenets for others to follow. Quality must be a personal thing; it requires personal involvement.

Ensuring Employee Buy In

Many recognize the merits of quality but aren't sure where or how to start. Doubts and concerns are easily identified. More often than not they will be similar to those listed in Figure 2.1. These were developed by a group of employees using a storyboarding technique described in Chapter Three. It is important to know these concerns will exist, and you shouldn't be surprised when they are exposed. In fact, I'd recommend getting them listed and made public very early. An important part of commitment is not to worry the quality subject to death but to proceed forthrightly. (During the same before-mentioned session, employees submitted their thoughts regarding the "positives" of quality (see Figure 2.2). Clearly the potential value in the pursuit was recognizable to them from the beginning.)

Before drafting the mission statement and objectives, time should be spent investigating what others have done and are doing. Today there is much more information and literature to draw from, and your course can be more easily charted than was true a few years ago.

Involve as many of your people as possible in the investigation-and-early-analysis phase because the broader the base of understanding and enthusiasm, the better. Strive to maximize the number of your people committed to quality early in the process. Some will enthusiastically endorse quality, others will more quietly go along with varying de-

QUESTIONS/CONCERNS
Employee Storyboarded Responses

• How will we have time to do it?

• What will be done with those not cooperating?

• Built-in resistance to outside suggestions/changes.

• What takes precedence, quality or dollars?

• Will this process be required of our contractors, suppliers, etc.?

• Concern with "Program of the Week" syndrome.

• Will there be a quality department?

• Dedication to resources?

• Do we really need this process?

• How do we improve employee attitudes?

• Union cooperation?

• How do we define quality?

• Is the quality effort cost effective?

• Is executive management totally committed and financially supportive?

• How do we respond to suggestions?

• Do we have enough staff to handle it?

• Will "It" affect our staff scheduling?

• How does "It" affect the budgeting function?

• How will "It" work?

• What happens to management who won't buy in?

FIGURE 2.1 Storyboarding can reveal doubts about the quality process.

POSITIVES
Employee Storyboarded Responses

- Recognizing and meeting customer needs (internal/external)
- Cooperative attitudes
- Identify opportunities
- Better planning
- Improve efficiency and productivity
- Better defined and understood company-wide goals
- Improved problem-solving tools
- Improved employee attitudes, morale, and recognition
- Benefits of internal customer concept
- Consistent (high) quality
- Improved communications
- Dollar savings

FIGURE 2.2 Storyboarding can also reinforce positive aspects of quality.

grees of skepticism, while a few will resist the change, perhaps even fervently. Buy in is extremely important, but it doesn't happen overnight: Therefore, you must give the impression that your commitment will last "forever"; the longer it is seen and felt, the stronger it becomes.

In the beginning commitment provides the foundation upon which to build. Later it becomes the quiet reassurance those weak of spirit need to venture into the quality arena. The change to quality is done a step and a person at a time. Causing early broad-based involvement in the development of the quality process can be evidence in itself of the new commitment. It has the potential of demonstrating immediately the difference quality can have upon how things are done.

Shakespeare wrote: "Our doubts are traitors and oft make us lose the good we might win, by fearing to attempt." Commit to the quality process, fear not, and go forward.

Chapter 3

Develop Your Quality Mission Statement and Objectives

IMPLEMENTATION STRATEGIES

The development of the quality mission statement and objectives should be a team effort. The first step in their development is to form an implementation team to draft the statements. Once the quality mission and objectives are drafted discuss them with a large discussion group before finalizing them. I believe that if the quality effort is to be implemented company-wide, it is best that the drafters be a mix of executives and senior-level management. The discussion group should consist of representatives from all levels of management. Although the intent will be to involve as much of the total work force as possible in the process, the development of the quality mission statement, objectives, process to be utilized, and plan for its corporate implementation is a management responsibility.

Departmental Approach

The implementation team also should be responsible for monitoring the progress of the quality implementation to determine where modifications should be made and how to improve the process. However, as the quality process spreads throughout the organization, individual departments and work groups should be encouraged to develop their own implementation strategies. This is an excellent way to capitalize upon the operational and demographic differences found within a large enterprise. Therefore there needs to be considerable flexibility in the implementation plan, yet a firm organization-wide commitment to the corporate quality mission statement and objectives being pursued, the overall plan for attainment, and the process to be utilized.

Implementation Plan

Figure 3.1 illustrates the rollout steps and the importance of the consistency required along the way. The structure, therefore, of the quality mission statement, objectives, process, and implementation plan must be recognizable to individuals as they become quality participants and begin using quality tools and techniques.

It is not uncommon for too little attention to be given to setting the stage for quality. Unfortunately, the individuals expected to participate and achieve the desired accomplishment, changes, and/or improvements are consequently left lacking and for the most part underutilized in the pursuit. Individuals must understand why the process is important and what is desired. Only then can the tools and techniques achieve their fullest potential. Too many times we train our employees in how to use quality tools without ever telling them what our mission statement and objec-

Corporate Quality Mission Statement and Objectives

↓

Corporate Quality Process

↓

Corporate Quality Implementation Plan

↓

Divisional Quality Implementation Plans Must:

- be consistent with the corporate
quality mission statement and objectives

- utilize the corporate quality process

- complement the corporate quality
implementation plan

↓

Departmental Quality Implementation Plans Must:

- be consistent with the corporate
quality mission statement and objectives

- utilize the corporate quality process

- complement the corporate and
divisional quality implementation plans

↓

Individuals Participating in Quality Implementation Must Know:

- the corporate quality mission statement
and objectives

- the corporate quality process

- the applicable quality implementation plan
they are pursuing

FIGURE 3.1 The quality-improvement implementation process.

tives are. Too often we expect individuals to accomplish their daily work tasks without ever telling them why it is being done. Therefore, it is not enough to *establish* the quality mission statement and objectives; they must be *understood* by every person expected to use the process.

NATIONAL QUALITY AWARD: BLUEPRINT FOR SUCCESS

In 1987, Congress passed Public Law 100-107 which established the Malcolm Baldrige National Quality Award, managed by the U.S. Department of Commerce's National Institute of Standards and Technology and administered by The Malcolm Baldrige National Quality Award Consortium, Inc.

Anyone who wants to improve quality should become familiar with the criteria the Malcolm Baldrige Award uses to determine quality achievement and excellence. The Award's seven examination categories, subcategories, and their point values define all elements comprising quality, and as the quality mission statement and objectives are being shaped, the Award criteria can be extremely useful (see Figure 3.2). Later they can be invaluable as an assessment guide for monitoring implementation progress and achievement.

QUALITY MISSION STATEMENTS

Our mission statement (see Figure 3.3) was written prior to the establishment of the Malcolm Baldrige Award but it contains the ingredients necessary for an organization-wide feeling about what is desired and how it is intended to be accomplished.

As important as mission statement and objectives are, even more important is the way they are developed. Word by word, after sometimes tediously long discussion and fre-

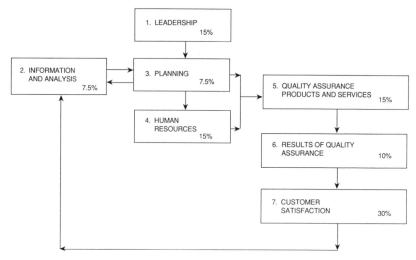

FIGURE 3.2 National Quality Award examination categories.

CITIZENS GAS & COKE UTILITY QUALITY PROCESS

Mission Statement:
The quality mission of Citizens Gas & Coke Utility is to satisfy the needs of our customers, both internal as well as external. Together we will create and maintain an environment where all are motivated toward the satisfaction of customer needs, each contributing his or her own unique talents and abilities to the process. Through mutual respect, personal pride and teamwork, we will strive for excellence – continually improving our services, processes, and products. We will work to be the best at what we do, in every phase of our business, at every level.

FIGURE 3.3 Criteria for quality: Organization-wide objectives.

quent debate, the implementation team formed our expression of intent. The result was a feeling of ownership, bringing with it individual buy in and commitment. This is essential from the beginning and contributes to greater involvement, which spreads first-person knowledge of the process throughout the worker grapevine.

Even though great similarity exists in most mission statements, don't succumb to the temptation of a cut-and-paste transferral to your own situation. Take the time to define your own guiding philosophy. The process of developing a mission statement provides a wonderful opportunity to demonstrate a new way of doing things by involving employees more than usual and is a useful vehicle for enlisting supporters from the beginning.

STORYBOARDING

Attention should be given to the method used to gather employee input, and I believe storyboarding, developed by Walt Disney, is a simple but extremely effective technique for enabling team members to express their ideas and recommendations quickly and anonymously. The team referred to here would most likely be your quality implementation team whose charter would be to establish the focus, mission statement, objectives, and implementation plan of your pursuit. (Team formation and conduct guidelines are discussed in Chapters 11 and 12.) The following describes the storyboarding technique:

- Provide several cards of the same color to each member along with felt-tip marking pens of the same color.
- Ask the question to be considered (e.g., "What should be the key elements of our quality mission?"). It expedites the process if the question has been previously stated by memo or listed on the agenda.

- Ask for responses to be phrased in two to four words, one idea to a card.
- Collect the cards after five minutes and begin posting them on a large cork board. Provide a few more minutes for team members to see the first cards being posted so additional thoughts can be generated.
- Arrange the cards into common groupings, discarding, with the team's permission, those that are identical.
- Ask the team to reach a consensus regarding the appropriate heading for each grouping of cards and use a different colored card to indicate so.

If additional discussion of the storyboard is required beyond the time allotted for the team meeting, a picture of the board may be taken so it can be recreated at the next meeting. It also is very helpful to distribute a list of the recommendations after the meeting.

Time is critical to the quality process, and storyboarding minimizes undesired discussion and debate of subjects. It also reduces the influence of individual team members and permits the airing of otherwise controversial interdepartmental or individual differences. Meetings should not exceed one-and-a-half hours.

In spite of its simplicity, storyboarding is an extremely effective technique for securing several ideas, thoughts, or feelings in a very short time frame. Of equal importance is that it not only provides the opportunity for all team members to participate, but it encourages their participation.

When initially used, storyboarding may be the first time that some group members talk about certain issues in front of their peers. Psychologically it assists more reserved group members to be more comfortable in a team setting. Storyboarding helps the usually timid to realize that their ideas are just as good as, if not better than, other team members. Storyboarding eliminates the threat of embarrassment, and future concerns about actively participating in team discussion diminish greatly.

I recommend the use of storyboarding at the first team meeting because early use creates an extensive "to do" or "to be considered" listing immediately and can jump-start the team in its quality pursuit.

Beyond team use, the technique is equally effective in a multitude of routine problem-solving settings. Storyboarding is so simple and effective that its broader use does not need to be encouraged, it simply happens. Almost overnight, it seemed, I saw cork boards, tripods, and storyboarding cards, pens, and pins appear in every meeting room and elsewhere. As I reflect upon this, I wonder how many ideas storyboarding has generated, particularly from those who previously had been silent.

DEVELOP OBJECTIVES

Once the team has drafted the mission statement, it should proceed immediately to development of the quality process objectives. The mission statement and objectives should be developed and viewed together, not separately, as quality is communicated to the organization.

As the team strives to reach a consensus on the mission statement, some of the ideas submitted will be more appropriately expressed as parts of the statement of objectives. In presenting the mission statement to the organization, it should be viewed as the spirit of the quality effort. The objectives should be seen as the specific targets for which pursuit strategies must be developed.

Again, as with the mission, the objectives must convey the corporate intent as to *what* the process is to achieve. Figure 3.4 illustrates objectives a team might develop. I believe the first objective should be: "To identify and satisfy customer needs."

QUALITY PROCESS OBJECTIVES

• To identify and satisfy customer needs

• To provide disciplined approaches to problem solving,
  team building, and decision making

• To achieve cost-effective operations by improving
  quality and eliminating waste

• To monitor quality and effectiveness through feedback

• To be better than our competitors in every area of
  our business

FIGURE 3.4 Develop objectives that convey the corporate intent.

The word "discipline" in the second objective listed in
Figure 3.4 should not be overlooked. As discussed in Chapter Two, the process requires organizational and individual
discipline in adhering to its mission statement, objectives,
and tenets.

MANAGEMENT BUY IN

Once the mission statement and objectives are completed,
they should be discussed by executive management to as-

sure and achieve group consensus. Don't expect unanimity for there will most certainly be one or two who can't or won't buy in to the change a quality culture creates. From the beginning, however, don't shrink from the sometimes louder voices of some who will be skeptical or obstructive. The challenge is to ascertain where the majority—the consensus—is. If it rests opposed to the concepts and practices of quality, then it must be reversed or else the effort is doomed from the start and an organizational disaster will occur. Recognize, too, that individual skepticism will be found every step of the way, even at times when the spirit is willing. Quality is a bold pursuit and sometimes requires a courage—or seemingly blind faith—some will have trouble finding.

Once executive management is comfortable with the mission statement and objectives, share them with the broadest possible managerial audience. Discuss them openly; storyboard the advantages and opportunities they suggest; and storyboard the concerns, questions, and/or doubts they evoke. Quality does not occur without open and honest communication. Usually two meetings will provide ample participation and input, but don't expect a consensus of the larger group. They'll have years of experience supporting the proof in their minds that executive management will never change to the degree quality requires. Statements of missions and objectives are only words—and everyone will remember failed programs of the past. Adherence to a process with a beginning but no end will be more than suspect.

Actions consistent with the stated mission and objectives provide the only convincing truth the vast majority of employees desire.

Chapter 4

Develop Your Quality Process

THE CUSTOMER-SATISFACTION PROCESS

A quality mission statement and objectives alone are not enough to cause an organization to adopt a quality culture. A process must be developed that can be utilized to achieve the desires of the mission statement and targets of the objectives.

Discussion—perhaps even debate—must occur to determine what elements are necessary for the successful implementation of the quality process. Admittedly, there are different approaches as to what quality should be. Each organization must find its own comfort level regarding the elements and options that define how quality is to be achieved.

As I stated earlier, I believe customers and their needs should drive the process, and no matter where you are in the methodology, you must be able to relate to your customers. Work processes should have a customer basis; so, too, should our plan strategies. Problem solving should, in the first step of analysis, determine customer impact. Measurements should have an identifiable customer meaning as well. In all that we do, we should know who we're doing it for and what the need, real or perceived, is.

The customer-satisfaction (i.e., quality) process I advocate consists of three phases of activity (see Figure 4.1), assessment, delivery, and monitoring for continual improvement. For maximum success, the customer-satisfaction process should be utilized by employees throughout the organization. Constant effort toward effective communication in support of the process is essential. Finally, a methodology must be present that permits disciplined resolution of problems, issues, and opportunities when they are encountered. These are the key elements of the customer-satisfaction process, and they may be pursued by teams or by an individual.

Assessment

Needs assessment must not be done in the vacuum of conventional or personal wisdom. There is only one way to determine customer needs and that is to establish a communicative linkage and ask customers to identify their needs—real or perceived. As we assess needs, we need to identify our customers, and we need to list our outputs. In other words, "What do we do?" and "Who do we do it for?"

The list of customer and outputs will be long and confusing at the outset. Don't worry about where to begin, however. It really doesn't matter where you start. Logically, it makes sense to prioritize the list, identifying the three or four most important customers or outputs. It may be best, however, to select a relatively easy customer or output at first to develop a sense of how the process works. In any case, remember quality is a process with a beginning but no end. It has no time constraint because once started it is never completed. Our long lists, therefore, also will be

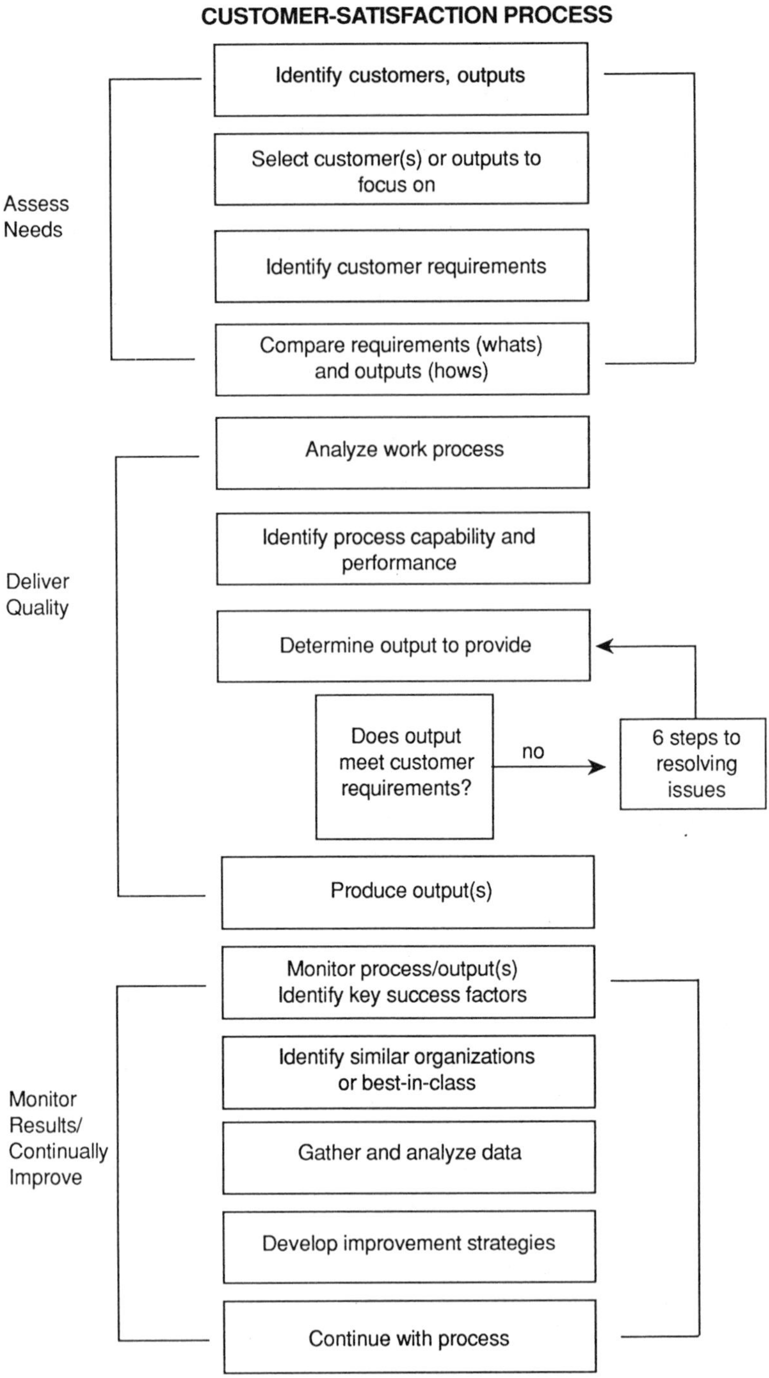

CUSTOMER-SATISFACTION PROCESS

FIGURE 4.1 The three phases of the customer-satisfaction process.

worked on by those who follow in later years. We are only the beginning.

Having selected the customer or output, next determine the actual customer requirement. You can begin by establishing what you think the requirement is, but in every case you must *ask* the customer to identify or confirm the customer's actual need.

Finally, compare the requirements (the "what" you are to provide) with your outputs (the "how" you will provide). Now, analyze your findings, looking for needs with no outputs or outputs with no needs. Consider, too, whether there are measurements for the requirements which can be used later to determine how well you are performing and/or to establish a comparative basis for improvement strategies. Figure 4.2 shows the activities of the assessment phase and questions to ponder as you proceed through it.

Asking the customer what *really* is required also should provide data useful in prioritizing your outputs. Don't be surprised to find outputs not nearly as critical as assumed or specifications established upon misinterpreted needs. In some instances the output may be what the customer desires, but even here we need to proceed into the other phases of the process, studying the delivery mechanism and related measurements, always striving for ways to improve.

Delivery

Figure 4.3 describes the second phase of the process, which concentrates on delivery, i.e., analyzing the work processes related to producing (providing) the output previously identified as required by the customer.

Study the various activities currently required to produce the output and consider their performance consistency and capability. It's important to contemplate this early in your pursuit whether you believe improvement opportunities exist or not.

Customer Satisfaction Process Guide
-Assess Needs -

ACTIVITY	QUESTIONS TO CONSIDER
Identify Customers, Outputs	Who are the customers? What do we provide?

> List them all – there is no
> right or wrong answer!

Select Customer(s) or Output(s) to Focus Upon	What criteria should be used? Who are the key customers? What are the key outputs? Do you have different types of customers? Have you discussed the results of this step with your team advisor?

> If you are just starting,
> pick one – save the tough
> ones for later.

Identify Customer Requirements	Do you really know what the customer wants? Are there other needs that the customer has not stated? Have you negotiated the requirements with the customer? Have you reviewed current and past data? Do your customers all have the same needs?

> Start with what you think is
> the customer's need, then
> ask the customer.

Compare Requirements (Whats) and Outputs (Hows)	Are there needs with no outputs? Are there outputs with no needs? Are there specifications or measures for the requirements? How does the customer feel you are doing? How does the customer rate you against the competition? How do you feel you are doing?

> Organize the data and
> do some analysis

FIGURE 4.2 First identify customer requirements.

Finally, determine the output your analysis indicates should be provided, comparing your recommendations to the identified customer requirement. There very well may be conflicts at this point because the output may not meet customer requirements, or the delivery capability may be

Customer Satisfaction Process Guide
- Deliver Quality -

ACTIVITY	QUESTIONS TO CONSIDER

Analyze Work Process

> Make sure you get into the details!

How do you produce the output?
Can you diagram it?
Are there other ways to do the job?
Do others do it the same way?
Do others do it differently?
Can it be simplified?
How do you know if you are doing a
   good job?

Identify Process Capability and Performance

> You may need more data for this.

What is the process capable of providing?
How consistent is the process?
Can you always predict the results of
   the work?

Determine Output to Provide

> Test a few ideas – maybe first idea isn't best.

Does the output meet customer
   requirements? If not, why not?
Have you discussed your results with your
   team advisor?

Produce Outputs
Make Changes

> Consider a pilot program or trial run.

If there are changes needed, do you have a
   plan outlining how you will get them done
   (who, what, and when)?
Do you have everyone involved who needs
   to know about the changes?

FIGURE 4.3 Analyze data and implement the quality process to meet those requirements.

restricted by procedures, standards, budgetary capabilities, or other obstacles. It is at this point we must force the consideration away from "why we can't" to "how we can." Discipline again must prevail, not only as a "can do" mind-set but in utilization of a systematic issue-resolution approach. We use the six-step methodology described in Figure 4.4 and strive to assure we spend sufficient time in identifying and analyzing the issue—or opportunity—and the customer-service impact (Steps 1 and 2).

Too often the tendency is to begin by generating solutions or, worse yet, to select a solution without actually having done an analysis of the issue and its impact. Upper management may be more prone to this "hurried up" fail-

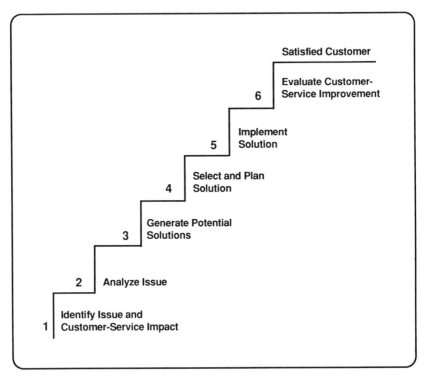

FIGURE 4.4 Six steps to resolving issues.

ure, convinced by their past experience—or current position—that they know what is best. When this happens we fail our organization in at least two ways: We haven't permitted proper analysis of the issue, and we haven't sought the ideas of others or *listened* to those offered. Obviously such failure seriously affects the attitudes of those around us as well as limits the power of combining more than one idea with a thorough analysis of our options.

Monitoring

When output-and-requirement resolution is reached (and this may mean some degree of compromise may have been negotiated), the next step is to proceed with providing the output and begin the third part of the process: monitoring results and striving for continual improvement (see Figure 4.5). Since measurements are the vital elements utilized at this point, we must establish the key factors to reflect how successfully the delivery mechanism performs and how well we satisfy the customer's needs.

When measurement is driven by customer needs, tremendous opportunity for improvement may be found. Too often measurements have been established to provide well-intended productivity references but without regard to satisfactorily fulfilling the needs of our customers. For example, measuring the number of customer calls handled per hour by a customer service representative reveals nothing about whether the reason or issue was satisfactorily resolved. The more cost-effective measurement should recognize that more time spent in meeting the customer's needs in the beginning may mean less resolution time required by the organization overall. We must have this broader cross-functional appreciation for how our actions impact others, which is nothing more than seeing other individuals and departments as our internal customers.

Measuring against our own performance and striving to

Customer Satisfaction Process Guide
- Monitor Results/Continually Improve -

ACTIVITY   **QUESTIONS TO CONSIDER**

Monitor Process/Outputs
Identify Key Factors

How do you measure performance?
Where is improvement needed?
What does the team advisor think?

You need to gather enough data to give a clear picture.

What should you monitor to guarantee that your product or service will continue to meet the needs of the customer?

Identify Competitors or
Best-in-Class

Who is the best in marketing, finance, cost, quality, and delivery?

First brainstorm possibilities, **then** narrow down.

Gather and Analyze Data

Can you improve the process or the output?
What have you learned from your competitors or others with similar processes?

Get into more detail – reexamine cause and effect relationships.

Develop Improvement Strategy

What is your improvement target?
When and how do you plan to do it?
What help or resources will you need?
What are the action steps and timetable?
Are you continuing to monitor the process?

Plan your work before working your plan!

Continue the Process

What is the next priority?
Should you go into more detail on the same issue?
Should you pick other processes or outputs?

What do you want to work on next?

FIGURE 4.5 Striving for continual improvement.

improve upon it will provide a meaningful reward. Even greater achievement may be attained by studying those who provide similar outputs or do similar tasks. Often linkages can be formed where measurement data is exchanged, not for the sake of simply comparing numbers to numbers, but, more importantly, to identify those who may be performing better than you. Once identified, the challenge is to determine *why* or *how* their delivery mechanism is different. From this analysis should come improvement strategies for better meeting the needs of customers and doing so more effectively.

It's important to keep the organizational focus upon the customer-satisfaction process and strive to make assessment, delivery of those needs, issue resolution, and measurement and continued improvement the routine way to go about your business rather than something extra you do or an afterthought. From the beginning, commit to be disciplined, almost unrelenting (in a quality sort of way) in use of the process.

Chapter 5

Develop Your Implementation Plan

Completion of the first phase of the quality pursuit is accomplished by developing a formalized implementation plan. This activity provides several opportunities to illustrate the use of the newly developed customer-satisfaction process and to instill a new mind-set regarding improved communications and employee involvement.

QUALITY IS NO SECRET

Many—perhaps most—organizations limit the number of employees who are aware of the organization's formal plans. Plans often are looked upon as secret documents, developed and known only by a few select people. One of the powerful potentials of the quality-improvement process is the opportunity to change this restriction by making the implementation plan much more broad-based, visible, and dynamic.

Figure 5.1 illustrates how an organization's planning process may appear *after* total quality has been achieved.

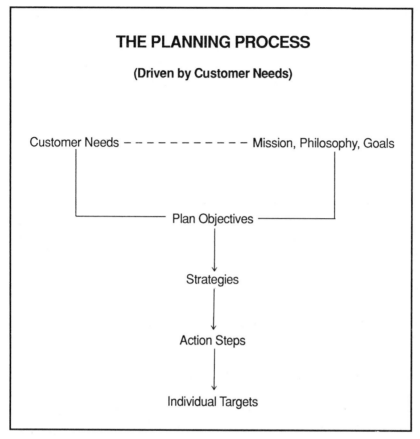

FIGURE 5.1 The planning process after total quality has been achieved.

(Total quality occurs when the use of the quality process and its techniques becomes the routine way tasks are performed. This may take five or six years of concentrated effort.)

CUSTOMER NEEDS DRIVE PLANNING

Customer needs influence our mission statement and goals, and are translated into plan objectives. For these objectives, departmental strategies are developed, each containing the various action steps required for accomplishment. The action steps can then become the targets for the individuals or groups of individuals (teams) to pursue. These, too, should have formal plans developed.

The effect of this planning cascade is to have the planning function driven by customer needs. It also enables employees to better understand why they're doing their various tasks.

Developing the quality implementation plan may be the first time various aspects of a new culture are directly or indirectly experienced. As previously stated, *use the Process to develop the plan* . . . i.e., assess the needs, develop the work processes (strategies) necessary to deliver (meet) those needs, and establish measurements (action steps) to determine when accomplishment is to occur. Monitoring of the delivery of needs also will identify additional needs and opportunities for further improvement. These, too, should be translated into formalized strategies.

THE QUALITY IMPLEMENTATION TEAM

The plan should be developed by your 10- to 12-member quality implementation team, which should be comprised of a cross-functional mix of executive and upper-level management.

At Citizens Gas & Coke Utility we debated whether a broader mix of employees, including union representation, is required on the implementation team. The way we proceeded worked fine, and we found that there is ample opportunity later on for involving others. Top management commitment is so vital that I suggest you not complicate the revolutionary idea of the first two levels of upper management working and communicating together.

Storyboarding the question "What do we need to have or do to implement quality?" will provide ample input for forming your first objectives, strategies, and action steps. We identified seventy-six different tasks necessary to initiate the effort. These were consolidated into six different strategy groupings under a single three-year objective (see Figure 5.2). These strategies are the first reiteration of the quality implementation plan and cannot be completed in the first year.

The organization strategy should include actions to:

- name the process;
- introduce the basic process concepts to upper and middle management;
- form an implementation team;
- develop and convey the implementation plan;
- determine staffing*;
- identify facilitation needs; and
- develop a budget.

*Note: I've found it best to keep the quality staff as small as possible. It is best if quality implementation is a routine responsibility of *every* department and is not viewed as a task a separate specific department does.

The recognition strategy should include actions to:

- develop team and individual recognition methodologies;
- consider management and other incentives;
- consider improvement opportunities in suggestion system; and
- implement appropriate recognition events.

The monitoring strategy should include actions to:

- identify and establish key performance indicators of the quality implementation plan;

FIRST YEAR QUALITY IMPLEMENTATION PLAN

Overall Objective: Within three years: implement organization-wide a formalized quality process.

Strategy 1:   (Organization) – Establish the quality process concept with upper and middle management and determine organization necessary to support its initiation.

Strategy 2:   (Education) – Educate all personnel in the new philosophy, process, tools, and techniques.

Strategy 3:   (Communication) – Implement ongoing communication strategies to maintain awareness about and progress of the process.

Strategy 4:   (Team Development) – Provide the opportunity for every employee to participate on a quality team.

Strategy 5:   (Recognition) – Implement recognition-and-rewards methodologies in support of the quality process.

Strategy 6:   (Monitoring) – Implement effective process monitoring-and-reporting methodology.

FIGURE 5.2   Six strategies for quality implementation.

- develop methodology to identify and correct implementation problems;
- develop methodology for internal and/or external assessments of implementation progress; and
- establish a methodology for developing improvement strategies.

These strategies will guide the initial efforts to implement quality. As the organization matures in its attitudes about quality, so, too, will the implementation plan. Many of the targeted actions stated simply in the beginning will require development of much broader strategies. The reality that quality is a never-ending, ongoing process, not started one day and finished the next, begins to be understood.

Part II

Laying the Foundation
Through Education

Chapter 6

Orient Your Employees

For quality to prosper, employees must understand more than the tasks, tools, and techniques. Unfortunately, often too little time is spent striving to convey the correct orientation for the pursuit. From the beginning, those leading the implementation effort must take the time to personally present *why* quality is to be implemented. Every effort should be made to make sure quality is not only understood but also *felt*. I believe this is further reason for the focus to be on the customer and satisfying customer needs. Customer satisfaction is something everyone can relate to, striving to achieve it has both tangible and intangible requirements.

The CEO and top management share the responsibility of establishing an appreciation for quality among all employees. This is a tremendous opportunity for top management to get to know some of the faceless names within the organization that can become living, breathing, feeling people sharing and discussing the organization's goals and needs.

Delegating these responsibilities will go a long way toward assuring failure. Explaining the "why" and "how" of the quality effort must be done by those recognized as being in charge. The more personal the explanation is, the better. It must be perceived as being a two-way communication with acknowledgment from the beginning of the scope of the commitment being undertaken. The approach must be gentle and understanding because change is a fearful thing, and employees need to be continually reassured of executive management's sensitivity to the realities of the journey.

The orientation sets the tone and establishes the style of the quality pursuit. Confidence in what will be accomplished and excitement for the quest can begin to provide the chemistry to make quality happen. These first messages must be repeated and restated time and again to constantly reinforce the purpose, but it is here we begin to communicate and to demonstrate the commitment and the involvement of those at the top. It is here employees will hear "I believe" and "Together . . . we . . . will strive to be . . . !" Quality needs to be emotional. And just as you may find it difficult to feel anything from reading these words, so, too, will employees if they are provided with only printed words.

The orientation should involve all employees in a common kickoff even though it will take time to educate and eventually directly involve everyone. The agenda of the orientation is to share the focus, mission, and objectives. Make sure the words and statements are felt. Try to inspire from the beginning but understand that probably more skepticism than belief may be the first reaction. Touch upon the process and the implementation plan, but ever so slightly. Remember, the process takes time and everything is not to be accomplished in a single meeting or at first blush.

Accept a certain amount of confusion in these beginning steps because they contain a multitude of new concepts and, in particular, new terminology. The language of quality sounds strange to the untrained ear. Yet its vocabulary has a hidden danger because many of its words have been commonly heard and are thought to be understood and perhaps believed to be used.

Expect only the slightest comprehension of what lies ahead. Orientation is at best only balancing the toddler in preparation for the first steps.

Chapter 7

Educate in Basic Concepts

ONGOING QUALITY EDUCATION

Part of the commitment to quality requires that *everyone* first be educated in the concepts and later in how to apply them. It is important to recognize the task is to educate, not simply train. Just as the quality process is ongoing, so, too, is the learning about it. Expect to develop a complete quality education curriculum over time. Also, reexamine existing training activities and assure they support and complement quality.

One of the early benefits of implementing quality can be in the classroom if the students are from cross-functional areas of the organization. For many, "working together" is at best within their own department, so the experience of learning together with fellow employees from different areas has great potential. Leave the organizational status stripes elsewhere, too. From the beginning, classes should be comprised of a mix of employees with a mix of responsibilities, from executive to line. Every step of the way, seize opportunities to convey the commitment and involvement of top management . . . the attitude of oneness in purpose . . . the spirit of teamwork.

QUALITY EDUCATION I

Quality Education I (QE1) class size should be kept small enough to be intimate (no more than thirty to a class). The objective of the first course should be to convey the basic quality concepts. Build upon the foundation laid in the orientation, repeating again the focus, mission, and objectives. A very effective way to start the class is to have the CEO give a refresher presentation on his or her earlier message.

Trainers should explain that the participants are their "customers," and the trainers' purpose is to be aware of and satisfy the students' needs. *Use* the process to educate about the process. Explain that after each presentation module, an assessment by the students is desired, including their rating of the effectiveness of the material and its delivery along with their comments and suggestions for improvement. In this way not only can course content be fine-tuned to better meet the needs of the students, but it also makes them active participants in the development of the class.

These evaluations should be seen as a "live" demonstration of the third phase of the process: measuring and monitoring for continued improvement.

The following outlines how you might construct a basic course in quality education (QE1) (information on sources mentioned is given in the Quality References section beginning on page 185). Realize, however, that only *you* can determine the course content best suited for your environment and objectives.

A. Opening
   1. *Quality Focus, Mission, and Objectives Statement by CEO*
      Seize this opportunity to illustrate your commitment to the quality pursuit. Employees expect important matters to have the backing and involvement of the person at the top. These initial words should convey your desire and intent to achieve a quality environment, not simply "support" the endeavor. This is your moment of truth—the time to state why quality is important to your organization.
   2. *Discussion of Mission Statement*
      The trainer should concentrate on making the quality mission statement come alive by discussing what the mission statement conveys and its implications for your organization.
   3. *Customer-Satisfaction Focus*
      Show the video "A Passion for Customers" by Tom Peters. This excellent motivational film presents examples of organizations that exceed customer needs and expectations. The goal of this segment is to develop within participants a feel for quality in the context of satisfying customer needs.
   4. *Teamwork/Customer-Needs Assessment Exercise*
      Conduct a hands-on exercise to demonstrate teamwork and, in particular, the importance of assessing customer needs by asking questions. This can be accomplished by dividing the class into small work groups to complete a specific task, such as building a bridge with tinker toys. Team members should be assigned varying tasks, from designing to building to inspecting. Team members must depend upon each other and communicate effectively to successfully complete their tasks.
      A detailed description of this exercise can be

found in *A Handbook of Structured Experiences for Human Relations Training,* edited by Pfeiffer and Jones.

Remember, the purpose of QE1 education is to establish a basic understanding of and appreciation for the concepts of quality. At this stage, you are merely laying the foundation for your quality endeavor.

B. Customer-Satisfaction Process

1. *Assessment of Needs*

   Teach participants to assess customer needs by asking questions and show them why this assessment is necessary to assure you are doing the right things. It is here you should ask "Who are our (your) customers?" "What are their needs?" "Do our (my) outputs satisfy their needs?" Storyboard responses to these questions, then discuss the responses. To complete the assessment phase of the course, show the video "I Know It When I See It."

2. *Delivery Process to Meet Needs*

   Discuss how "needs" require functional process or work systems to produce the desired output. Be careful to keep this first introduction simple. Relate it to examples relative to your organization or to generic examples that can be easily grasped.

   Discuss how to analyze, document, and measure a work process. Refer to concepts relative to input, output, and noise variables but don't get too technical. Flow chart a work process and identify and select the key measurements. Identify the output of the process and also show how to determine if it is the desired output. ("Does it meet the customer's requirement?")

   Show the video "Roadmap for Change—The

Deming Approach" to introduce Deming's 14 points.

Conduct the "Bead Box Exercise" to illustrate concepts of work process inspection and impacts of variation upon the work process.

3. *Monitoring/Measuring for Continual Improvement*
   In this portion of the course, stress the importance of recognizing that tasks are not completed when outputs are produced. The dynamic individual or organization will be constantly looking for ways to improve outputs, products, or services. This requires regular monitoring and measuring to know where you've been and where you are now. Within this context introduce the concept of "benchmarking"—comparing your measurements to those of your toughest competition or to the best performer of the task.

   Introduce to the class the concept of continual improvement being everyone's responsibility. Refer back to the Peters video to illustrate that truly outstanding organizations exceed customer needs and expectations.

C. Problem-Solving/Issue-Resolution Methodology

   Introduce issue-resolution methodology (Figure 4.4) in the context of making work processes satisfy customer needs.

   Stress the importance of developing the self-discipline to properly analyze issues by completing the proper steps. Also, reinforce that the customer impact consideration assures that we're addressing those problems, issues, or opportunities having the most significance.

   For each step, introduce techniques that support the classroom activities. Don't spend too much time on these techniques, however. The idea is to merely acquaint participants with the terminology and basic methodology of quality. The goal for this section of the course is for the class to know about the techniques that support the quality process and also know *when* these techniques *could be used.*

Step	*Activity*	*Techniques*
1	Identify issue & customer impact	Storyboard Brainstorm
2	Analyze issue	Cause & Effect Histogram Pareto
3	Generate potential solutions	Storyboard Brainstorm Interview Survey
4	Select and plan solution	Storyboard Cost benefit Balance sheet Gantt chart Pert chart
5	Implement solution	Establish monitoring measurements
6	Evaluate customer Service improvement	Measure Benchmark Verify with customer

D. Teamwork/Employee Involvement

Discuss desired cultural change and challenges. Storyboard obstacles. Show "The One-Minute Manager" by Ken Blanchard or show "Working Together Works." Outline team formation mechanics (family and cross-functional, reference Chapters 11 and 12) and team structure (leaders and facilitators, reference Chapter 8).

Conduct *Desert Survival and Subarctic Survival Situations* as individuals and then as teams to illustrate the power of teams.

E. Wrap Up

Review your mission, objectives, and implementation plan. Summarize the process methodology and techniques introduced.

Emphasize the importance of teamwork. Stress that this is only the beginning and while the implementation of quality takes time, it will occur if everyone in the organization works together.

Finally, conduct a course evaluation and solicit suggestions from the participants.

Remember, the intent of QE1 is to provide an introduction to the basic concepts of quality and teamwork. The participants should not be viewed as having been "trained" and therefore ready to go forth and "do" quality. They should, however, have a basic understanding of the direction (focus) of the effort, its methodology, and some helpful tools and techniques.

Be careful not to emphasize statistical techniques too heavily. These can be frightening and should be gently brought into the process as it matures. Those who have an aptitude for statistical analysis can be used later to assist teams as facilitators or internal consultants. This will increase the comfort level of others and, more importantly, apply the techniques to real-life situations rather than concocted classroom illustrations.

Other hints regarding how to best present QE1 are as follows:

• Use videos, break-out sessions, and exercises to keep the course interesting and lively

- Minimize lectures
- Provide written materials, including a course outline
- Avoid supervisor/employee match-ups
- Be sensitive to reading/writing capabilities
- Make illustrations pertinent to the audience (e.g., family or community examples)
- Avoid difficult to grasp terminology (e.g., statistical/ technical references)
- Avoid productivity-improvement implications (They are sometimes perceived as job threatening)
- Illustrate and support the process using break-out sessions and exercises
- Recap periodically
- Frequently illustrate the process methodology (Figure 4.1)
- *Have fun* while learning

For the first classes, use members of the quality implementation team as class monitors, considering ways to improve course content and delivery. This greatly assists in establishing buy-in and clearly demonstrates that the desired effective results go well beyond those of the Human Resources Department.

Evaluation of the course should examine the relevance of the material, its delivery, the quality of visuals, and the general interest of the presentation. For each of these, comments should be solicited and later seriously analyzed. Be careful not to let single comments weigh too heavily in your examination. Look instead for similar commentary and consider whether the thought expressed is a consensus view.

Chapter 8

Train Leaders and Facilitators

THE LEADER/FACILITATOR WORKSHOP

Quality teams (described in more detail in later chapters) provide the vehicle for applying the process. Successful team activity requires people trained in a variety of team-support techniques. Consequently, at the same time as QE1 is being presented, a second course, QE2, designed for quality leaders and facilitators, should be offered to those who will be supporting teams.

The purpose of the Leaders/Facilitators Workshop (QE2) is:

- to provide tools and techniques for conducting successful meetings
- to demonstrate how leaders and facilitators can assist their teams
- to discuss the roles of others involved with the team
- to establish a support network of employees promoting the process and sharing experiences with each other

I recommend all exempt employees attend QE2 because logically they are the ones expected to lead and support the process. However, the workshop should be open to anyone who is interested whether they are likely to be team leaders or facilitators. The fact is that anyone attending QE2 will be a better team member.

The following outlines how QE2 (quality education 2) might be structured. While QE1 is *education* in basic concepts, QE2 is *training* in specific techniques. Again, you must design your course to complement your quality pursuit (information on sources mentioned is given in the Quality References section beginning on page 185). Participants taking QE2 should already have completed QE1.

A. *Introduction*
  1. Explain QE2 Purpose
       Explain the purpose of the QE 2 course. Remember to explain why education in quality is being provided and what the desired results are. Though this may seem basic, we often fail to provide a rationale for learning and consequently end up teaching "how to" without a foundation in the basics.
       It is extremely important to link learning experiences. Linking connects what otherwise might appear to be random, unrelated teachings. Linking also reminds participants (employees) of what they have learned in previous classes and reinforces the substantive nature of our total quality pursuit.
  2. A good opening for QE2 is the video "Brainpower" by John Houseman.
  3. Review QE1
       a. Mission and objectives of total quality pursuit.
            Refer to your quality mission statement

and its objectives at every opportunity—at every educational/training experience, initial team meeting, department meeting, and employee gathering. This reinforces your organization's quality foundation.

b. Customer Satisfaction Process

Review the "Assess, Deliver, Monitor" methodology described in Chapter 4.

c. Monitoring (Measuring) Techniques

In QE1, techniques that can be used to support the quality process methodology are introduced. It is important to provide QE2 participants with more in-depth exposure to these techniques so they can later use them to support team pursuits.

Chapter 9 refers to "The Memory Jogger" as an excellent reference on quality tools and techniques. In QE2, an in-depth review of this booklet as a training aid is suggested. Link this review to the "Monitor results/Continually improve" phase of the quality process and also to the "Six steps to resolving issues" (See Figures 4.4 and 4.5). This will provide both awareness and appreciation for where to use these techniques as well as how to use them.

d. Six Steps to Resolving Issues

Review the "six steps" described in Chapter 4. Remind participants that a disciplined use of quality techniques and methodology will occur only if we remind ourselves of, repeat, and demonstrate these techniques.

e. Teamwork/Employee Involvement

Remind participants of the effectiveness of working together using experiences shared in QE1. Emphasize, for example, the absence of the word "I" in the mission statement and the presence of the words "we" and "together."

Demonstrate to QE2 participants your intention to delegate—to empower a broad number of employees in the implementation of quality. Here, too, is an opportunity to steer employee attitudes away from the traditional "career-ladder advancement" concept to that of cross-functional support and leadership. Show "Team Building."

B. *Describe/Define Teams and Team Activity*

1. Discuss the types and purposes of "family" or "cross functional" teams (See Chapters 11 and 12).
2. Discuss Composition and Mechanics of Teams

    Be certain participants have a clear picture of how teams are formed, as well as their size, and function. Include guidelines on how to develop and submit recommendations. (These guidelines should be developed by your implementation team and should include criteria compatible with your organization. Be sure to emphasize the importance of timely responses to recommendations.)

C. *Describe the Roles of:*
    - Advisor
    - Leader
    - Facilitator
    - Recorder
    - Participants

1. Storyboard Exercise: "Coming into this class, what concerns do you have about being a team facilitator? a team leader? a team participant?
    (Besides providing another opportunity to use storyboarding, this exercise demonstrates that class participants share common concerns.)

D. *Discuss "How To Get Team Activity Started"*

    The appendix provides a sample outline for getting a team started. Something like this should be given to class participants as a reference source.

E. *Communication*

Conclude QE2 by showing and discussing the video "Dealing with Difficult People from Phoenix." Review other communication issues pertinent to team activity and answer related questions. Have participants evaluate the QE2 course.

SUPPORTING THE TEAM EFFORT

Team formation and related issues are discussed in Chapters Eight and Nine. The balance of this chapter describes the roles of the team advisor, leader, and facilitator.

Team Advisor

The team advisor is responsible for developing and communicating the team charter so it is clearly understood by all team members. The advisor should provide hands-on assistance in the beginning, but this should lessen as the team matures. The team advisor is usually a member of management and generally will be the person having supervisory responsibility for the pursuit described in the team charter. This person must be careful not to inflict his or her will upon the team's efforts but should remain close enough to the effort to assist the leader and facilitator in keeping them aware of policy considerations and in resolving team conflicts. Usually this can be accomplished by reviewing team minutes with the leader and/or facilitator.

The advisor must be careful not to be overly aloof or removed from the workings of the team. It is a natural tendency to think teams should be left alone—and to a degree they should—but guidance and support are still required.

The advisor's primary responsibility is to respond to the teams' recommendations. I strongly encourage making an event out of each team's presentation of its final report. Invite other teams, departments, and/or management, thereby emphasizing the importance of the team's efforts and publicizing the results. The advisor must assure, however, that responses to the recommendations are given to all team members, and if time for consideration is required, anticipated response target dates must be conveyed.

Team Leader

The team leader is responsible for assuring the effective operation of the team consistent with its charter. Generally, the leader should *not* have supervisory authority over any team member. The basic responsibilities of the leader are as follows:

- Schedule and arrange for and conduct meetings
- Prepare and distribute agenda before the meeting (This should include pre-meeting consideration of topics to be discussed and applicable research of them.)
- Assure agenda is followed during meeting
- Assure quality process and techniques are utilized
- Encourage involvement of all team members
- Keep team advisor appraised of status
- Assure team minutes are properly prepared by recorder and distributed
- Coordinate development of recommendations and their preparation

The team leader should be the communications hub of the team, before, during, and after team meetings. During the meeting the leader must be actively involved. Questions rather than statements should be the leader's participative style; he or she may find it necessary to have one-on-one discussions with team members to achieve team harmony and/or success.

Team Facilitator

The team facilitator is like a process-and-techniques consultant to the team. The facilitator is responsible for assuring the team adheres to its agenda, uses the process, and follows correct team-meeting conduct. Usually this will be best accomplished by being a listener during team discussions rather than an active participant. There will be times, however, when the facilitator should enter into the discussion to help the team ponder its conduct or activity.
Other facilitator responsibilities include:

- Assure leader is using process and quality techniques
- Assist, guide, and coach teams in use of the process and quality techniques
- Use questions to challenge the group
- Suggest appropriate tools and techniques to help the team identify or analyze issues
- Coach leader *after the meeting*, recommending ways to improve team activity

The facilitator also provides the linkage from the team to the point of overall coordination of organization-wide quality implementation, maintaining an awareness of team progress and identifying additional team needs. Facilitators should be looked upon as being the quality professionals within the organization, demonstrating by example. Often they will have had team leader experience, although never at the same time for the same team.
The QE2 workshop provides instruction as well for team recorders and members in general. Each team should have a recorder responsible for maintaining minutes and for writing down storyboarding and team thoughts during the meeting. The class also should provide refresher instruction from QE1 in the responsibilities of team membership with emphasis upon:

- Preparation *prior* to the meeting
- Participation during the meeting
- Acceptance of consensus findings
- Characteristics of a good team member

It is crucial to the success of the team that its leader and facilitator have had the additional training provided by QE2. Quality cannot be expected to flourish without providing for its infrastructural needs. Even in meeting this need there can be found tremendous opportunity for organizational improvement as additional employees are trained in leadership and participatory skills.

Chapter 9

Provide Basic Tools and Techniques

This may be one of the more important chapters of this book because we often fail to realize we need time to learn how to best use the quality tools and techniques. For maximum effectiveness, quality techniques must not be viewed as ends in themselves but should be accepted as aids to support the quality pursuit.

TEAMS MAKE THE BEST CLASSROOMS

The best classroom for most of us is to see the practical application of decision-making tools in a real setting rather than in abstract examples. Leaders and facilitators should learn the techniques in QE2 and demonstrate and teach as they proceed with assisting a team toward using the quality process to satisfy its charter.

Eventually everyone should be exposed to the following concepts:

- Flow Charting—cause-and-effect diagram
- Pareto Analysis—histograms and pie charts
- Cost-Benefit Analysis—pro-con (balance) sheet analysis
- Gantt Chart—critical path analysis

By seeing these techniques applied in the team setting, individuals may, over time, find them equally useful in supporting individual task improvement.

An extremely useful pocket guide to several of these techniques is called "The Memory Jogger." I suggest a copy be given to each leader and facilitator for quick reference and to assist in their application of graphical problem-solving techniques as appropriate.

KNOW WHAT TO MEASURE

Don't rush the measurement activity. The fact is we may have been measuring the wrong things in the name of productivity but at the expense of satisfying customer needs. For example, measuring the number of customer calls a customer service representative handles each hour ignores the concept of *satisfactorily handling* customer calls, which is the more cost-effective focus. We send entirely the wrong message to our employees—and to our customers—if satisfying customer needs isn't in the forefront. Likewise, in a manufacturing setting using statistical process control to develop conformance awareness of a standard the customer doesn't care about does nothing more than provide good information about the wrong thing. In every case we should strive to know first *what* we should measure before considering *how* to measure it best.

ALTERNATIVE TOOLS

Although some will argue the point, I believe the most beneficial tools and techniques for *early* support of a quality process are not those we traditionally offer. Instead, in the beginning concentrate on some or all of the following:

Communication

A quality environment is dependent upon excellent communication and, in particular, the individual's ability to do so. Techniques should be offered to assist in improving oral skills in one-on-one settings (including how to listen), as well as for presentations. A better understanding of the nonverbal messages sent via facial expression and/or body language is important, too. Also provide the opportunity to improve written skills for note taking, minutes, reports, memos, etc. Recognize that we all need to communicate and show how we can be better at it.

Team Work

Concepts in support of creating a team environment shouldn't be left to chance. Employees need to understand that often "two heads are better than one," and that teamwork doesn't require several people but can be as basic as two people working effectively *together*. The best teamwork techniques involve communication and work to change *attitudes* about individual involvement and responsibility. Quality is not something "they" or "the other guy" makes happen. It is something "I" do; something "we" make happen.

Storyboarding

Although storyboarding has been previously described (Chapter Three), it deserves mention here as a basic tool with significant influence upon implementing quality. Its value to team leaders and facilitators toward increasing participation and ideas cannot be overstated, but it can be effectively used beyond the formal team setting as well. The mechanics and variations of storyboarding should, therefore, be offered as part of your early training.

Interview/Surveying

Inasmuch as the process I advocate is dependent upon identifying customer needs, everyone should have a basic understanding of *how* to go about customer needs assessment. Remember, customers are internal as well as external, so there is a potential benefit in having everyone versed in how to go beyond his or her own interpretations of what the customers' needs are.

Many organizations already have training modules related to aspects of these techniques, but we have traditionally failed to provide the learning experience *broadly*. We assume, for example, that top management knows how to communicate, but the truth is we may not be very well versed in the mechanics at all.

Part III

Applying the Process

Chapter 10

Use the Quality Process

In Chapter Four I presented the action step "Develop Your Quality Process." Obviously it is not sufficient to stop with development. You must *use* the process and herein lies one of the fundamental truths of implementing quality: in order for it to happen, there must be a methodology that can be followed systematically by everyone in the organization. The process provides that methodology, and through it quality tools and techniques become meaningful.

USE THE PROCESS!

Too many organizations have no methodology—no way to get tasks done and improve upon them. The quality process applies a common target—satisfying customer needs—to everything done, but, again, the process must be more than words. It must be used. The first application steps occur in two different ways. One is more classroom education, the other is actual use of the process.

A basic course in quality, such as QE1, should not be

expected to prepare the individual for daily use of quality concepts and techniques. This will occur only over time and through experience . . . *provided* we continually remind and reinforce.

Applying the Customer-Satisfaction Process

An excellent approach is to develop a third quality course to:

- develop a better understanding of the quality process
- show how individuals can use the process to improve their output
- expand the monitoring (third) phase of the process and show how it produces continuous improvement

Note that these objectives bear a strong resemblance to productivity-improvement elements. However, I remind you that because the focus is first and foremost upon the *customer,* improvement ceases to be something management wants (in order to make more money) and becomes what we all should do (to better meet the needs of our customers).

Better understanding of the process is two-sided. One is from the perspective of the individual who through reiteration and new examples will begin to grasp that which may have been previously missed. Thinking and feeling about quality also requires developing a vocabulary and for this too, QE3 provides a refresher.

A follow-up course has another advantage, too, because some amount of time will have elapsed during which the employee will have first heard of quality in an orientation session; been introduced to its concepts in QE1; been either involved in or aware of a leader/facilitator QE2; and involved in or aware of others involved in team activity. The net of this time elapse is a general acceptance that quality is to be pursued on an ongoing basis and is not a short-term project or program. In other words, over time the individual becomes less skeptical and more receptive. Consequently, buy in becomes easier, and walls of doubt collapse.

Use the Process to Improve the Process

The other side of "better understanding" comes from those coordinating and orchestrating the rollout of the process. In the beginning, the presentation of quality is mostly top-down, and, even though employees' reactions are measured and suggestions are solicited, it is not *driven* by employee needs. Time changes this and the use of the process to improve the process becomes more clearly in evidence. Assessment of employee needs (internal customers) should cause examination of how the needs are being met (delivery phase of the process), and measuring and monitoring for continual improvement (the third quality phase) should provide the strategies for QE3 course content and related pursuits.

As stated in the beginning of this chapter, you must remind and reinforce over time to establish a true understanding of the process. A QE3 will provide the opportunity to emphasize that the process is applicable not only to team use but also to individual use. It is at this point I have found an organization can begin its continual improvement education. Call this course "Quality Education Three—Applying the Customer-Satisfaction Process" (with emphasis upon "Applying"). (Remember from the beginning I defined quality as "satisfying internal and external customer needs," so in our vocabulary quality and customer satisfaction are synonymous.)

Course Outline

Your course content might be as follows:

A. Introduction to QE3

   1. Opening comments by the CEO

     Give a "progress report" on how the implementation of quality is going. Talk about the number of teams formed to date, employee involvement, and the most significant achievements.

     Remind participants why quality is important to you and the organization and reemphasize your personal commitment to the effort. Enough time in your organization's pursuit has elapsed to make your restatement of commitment significant—providing further evidence that quality is not a program or project that will go away.

     In your comments, stress the importance of continual improvement in meeting customer needs and individual responsibility for improving job tasks. You can now begin to shift emphasis from the process to improvement, results, and higher expectations for achievement

     The foundation for quality should be in place by the time QE3 occurs; therefore, it is time to encourage the organization to stretch.

   2. Discuss QE3 goals

     Always be sure class participants understand the objectives for their study. In this course, these are to:

- Provide a refresher about the quality process and its methodology.
- Demonstrate the applicability of the process to individual as well as team efforts.
- Sharpen the organization's focus upon continual improvement as the responsibility of each individual and as the way to add value to products and services.

   3. Review your quality mission statement and objectives

Always begin education or training with this review to reinforce the basic principles of your pursuit.

Your mission statement and objectives also provide the common thread—the linkage—between all your developmental efforts.

B. Refresher on the customer satisfaction process and the six-step issue resolution methodologies.

1. Review process

Review the quality process, phase by phase, step by step (See Chapter 4).

Emphasize not only what needs to be done along the way but how to do it. Use past experiences of your organization in applying the process as specific references to make the methodology come to life.

Note: A very good lead-in to your review is the segment about Stew Leonard's grocery in Norwalk, Conn. from "In Search of Excellence."

This provides an excellent tone and focus for a discussion on satisfying customers needs and expectations.

2. Demonstrate use of the process

From the class participants, select two different types of jobs and illustrate the steps of the process. This takes the methodology out of abstract terms into real-life applicability.

C. Tools and Techniques

Discuss and illustrate process monitoring (measuring) techniques. (See Chapter 8.) The majority of your organization should be ready to grasp the applicability and importance of basic quality assurance and improvement techniques.

Use Chapter 20 to reference where each can be effectively used.

D. Continual Improvement (Reference Quality Process Phase III)

1. Discuss concepts shared in Chapter 22

   Try not to "lecture" about these concepts but rather work toward achieving participants' acceptance of them. Again, your goal should be to have each individual accept personal responsibility for improvement.

2. Reflect on the impact of change

   Show the video "Discovering the Future: The Business of Paradigms" by Joel Barker, and discuss the paradigms that exist in your organization.

3. Show the "Johnsville Sausage" excerpt from "Tom Peters: The Leadership Alliance."

   (This provides a "you can do more than you think you can" wrap-up for QE3. It challenges traditional concepts of management, so preview it to determine whether your organization is ready to wrestle with empowerment issues.)

Complement these topics with examples from your own organization of team pursuits and experiences. Include acknowledgment of occasional frustration or confusion and, in particular, the fact that the effort is ongoing and takes time. The point here is to be reassuring because each of us at one time or another will find the quality road a little bumpy.

QUALITY AWARENESS BEYOND THE CLASSROOM

The greatest reassurances come, however, not from the classroom but from *seeing* quality pursued. Opportunities abound to do this but at first they may not be clearly visible. Look for existing committees or work groups handling special challenges or strategic issues and have them incorporate the quality process and techniques into their efforts. The main thing is to *use the process*. This will seem forced or artificial in the beginning, but it is part of becoming more disciplined in the way we do things. As I look back, I continue to be stunned by the realization of how much time has been wasted and how many opportunities were missed by not having had a process to follow or the discipline to use it.

Graphically depict the process phases, steps, and techniques (See Figures 4.1 and 4.4). Make 8½" by 11" stand-up desk sets of these so every desk has upon it the constant reminder to assess, deliver, and monitor, as well as the six steps to resolving issues. Also, provide pocket and wallet cards so all employees can carry the guides with them.

Instill in the process champions, leaders, and facilitators an appreciation of the powerful example they can be and encourage them to incorporate quality tools and techniques into the way they go about their daily tasks. A new way of thinking, talking, and doing will be seen and felt far beyond the workplace where first used.

The transformation to a total quality culture occurs a step at a time . . . a person at a time. Be patient in your desire to get there but be unrelenting in your efforts. Although at the time of execution progress may seem painfully slow, you'll be amazed at how much change will occur over a relatively short period of time. My experience is that by the third year, the pursuit will have a momentum of its own and much less pushing will be required. In fact, the early champions may be challenged to keep pace with the change taking place.

But from the beginning, be disciplined in using the process. Keep it in front of everyone, talk about it, ask questions about it, promote its use, acknowledge those who have used it, and convey their use to others. In short, make the process real.

Chapter 11

Form Family Teams

Teams provide the vehicle for using the process to satisfy customer needs and to involve employees. I describe teams as being of two types: family and cross-functional.

Cross-functional teams will be described in the next chapter, but basically they are comprised of individuals from more than one functional area who have been brought together to address issues of a broad nature.

Family teams are comprised of individuals of a common functional area who meet periodically to apply the quality process tools and techniques to their daily tasks.

There will be a tendency in the beginning to think the pursuit is to form teams (i.e., promote teamwork) and/or to solve problems. Steadily establish the understanding that teams are not ends in themselves, they are simply the *means* . . . the way . . . to apply the process.

TEAM FORMATION

All teams should be formed by supervision to ensure they are properly chartered and have an advisor to coach them as they proceed. This will ensure the timing is right for team formation and that resources necessary for the team to function will be provided. The family teams I advocate are not, therefore, quality circles but are cooperative ventures where management and employees join together in a common effort. Family teams are the "together" of the "together we will build an environment" of the mission statement.

I suggest the following guidelines for team formation:

1. Teams should be comprised of individuals who have completed the basic quality education course (QE1).
2. A team should be formed only when a specific objective can be identified and chartered. In the beginning the objective probably will be to assess internal or external customer needs and through this issues/problems may be identified, some of which also may require team formation.
3. In these first phases of implementing the quality process, team formation should proceed slowly to assure proper management and facilitator support.
4. Family teams are comprised of individuals from the same work area and should be formed by a department head, manager, or supervisor.
   a. The team objective must be within the work and budgetary capability (parameters) of the area.
5. Cross-functional teams are comprised of individuals from two or more uncommon work areas; formation may occur by:
   a. Supervisor/manager/department head recognition that the team objective exceeds the capability of a family team and consequently other area(s) need to be involved.
   b. Executive management's desire to address a specific objective.
6. Whenever a team is formed, its formation should be communicated to the quality coordinator.

a. Any membership changes or role changes after formation also should be conveyed.
7. Team progress toward achieving its mission should be recorded. On a monthly basis, team leaders should provide an update of their progress to the quality coordinator.

Supervisors should be encouraged to enhance and expedite the process by forming family teams as soon as most individuals within the work group (family) have completed the basic QE1 course. Team formation should be recognized as the way structure is provided to the process. Through teams, ideas can be shared, new techniques learned, issues identified and resolved, and area objectives and strategies communicated.

QUALITY MUST COME FROM THE TOP DOWN

Quality should not be delegated; therefore, it should evolve from the top down in a cascade from one organizational level to the next. The CEO should have his own family team using the process to identify customer needs and formulate strategies for continually improving how needs are to be satisfied. Each of the CEO's team members should have his or her own family team and so forth down through the organization. This enhances management's understanding of process mechanics and provides an appreciation for team needs. It goes a long way toward achieving buy in as well as establishes credibility for the quality process among employees.

Figure 11.1 lists the benefits experienced through team activity. There also will be problems to be addressed as listed in Figure 11.2. These need to be anticipated and, to the extent possible, accepted as natural occurrences. In particular, be sure leaders and facilitators have been coached in how to recognize and resolve these problems. While these

STORYBOARDING EXERCISE

Team Benefits

What benefits or positives have you experienced in team activity?

- Process important
- Determining outputs, customers, and customer needs
- New mind-set to problem-solving
- Improved self-worth
- Positive attitudes
- Higher morale
- Improved communications
- Employee involvement
- Employees are more motivated
- New attitudes of team members back on the job
- Acceptance of responsibility
- Many individual ideas
- Working together (team work)
- Spirit of team togetherness
- Group buy in
- Consensus
- Improved employee input in solving problems
- Fixed the causes instead of the symptoms
- Improved listening skills
- More open atmosphere

FIGURE 11.1 Typical responses from a storyboarding exercise on the benefits of team activities.

STORYBOARDING EXERCISES

What problems or disappointments have you experienced?

• Resistance to change

• Some members want to use a different process

• Middle management's lack of conviction

• Attendance is sometimes a problem

• Not enough time for regular work and teams

• People who don't want change

• Straying from subject at hand

• Not everyone performs his or her role

• Inexperience with the process

• Clearly defining teams scope and mission

• Results are not expected for three to five years

• People may expect too much too early in the series of meetings, i.e., things will go faster

• Level of literacy of employees

• Quiet member

• Difficulty in leaving "hats at the door"

• Team's recommendation rejected by management

• Goal was too broad

• Nonproductive meetings leading to team disinterest

FIGURE 11.2 Typical responses from a storyboarding exercise on problems encountered in team activities.

problems may vary, most will relate to inadequacies in chartering and/or advising teams. Figures 11.3, 11.4, and 11.5 provide guidelines for advisors in chartering, maintaining, and closing out teams. Figure 11.6 provides a list of ques-

I. Forming a Team

A.   Draft the Team Charter

1.   State team advisor or person designated to coach the team concerning their project; to advise team regarding company policies and procedures; to offer recommendations or opinions as a guide to team action; to coordinate receipt of recommendations and provide implementation status information to team.

2.   Define the subject to be addressed, including key questions that should be answered. Also, include items or areas that should not be addressed.

3.   Identify the deliverables expected from the team. Examples include: objectives/ strategies, recommendations, analyses, etc. Indicate if deliverables will be in report form and/or a presentation, as well as to whom the deliverable will be submitted. (See Section III, Closing Out a Team.)

4.   Will the team be using the Customer Satisfaction Process, Six Steps or other process? Look for opportunities for measurements.

5.   Consider requiring that the team develop a work plan (action plan, goals, critical path network, Gantt chart, etc.)

6.   Consider requiring that the team keep a milestone chart – one or two significant items accomplished each meeting. Keep track of improvements/savings/results, if applicable.

7.   Indicate who should be copied on meeting minutes. Please copy the Quality Process Manager.

8.   State expected frequency of periodic update meetings (when the advisor will meet with the team leader and facilitator).

9.   Indicate a deadline date for the project and interim steps.

B.   Select potential team members and designate a leader. Call the Quality Process Manager for a facilitator (or have the leader call).

C.   Meet with the team leader and facilitator to discuss team charter, membership, etc. You should attend the first meeting to answer any questions that the team might have. (This is very helpful in getting the team off to a good start.)

D.   Select team members after discussion with the leader and facilitator, designate a recorder, and finalize the charter.

E.   Send "Team Formation Notification" to the Quality Process Manager and attach your charter.

FIGURE 11.3 Procedure for chartering the team.

II. Maintaining a Team

A.   Show genuine interest in the team's activities and accomplishments without "running" the team meetings.

     - Periodic planned visits to team meetings
     - Encouraging conversations with individual team members
     - Interest in Team Milestone Chart

B.   Meet with Leader and Facilitator on a scheduled basis to discuss the team's progress toward deadlines.

C.   Coach the team on laying out a work plan to accomplish its mission.

D.   Review team minutes and milestone chart.

E.   Monitor team leadership. (Some teams rotate leader, facilitator, and recorder functions periodically.)

F.   Assist members whose work schedule prevents them from attending meetings.

G.   Help the team stay on track with respect to budget and policies/procedures.

H.   Show interest and recognize individuals' ongoing team involvements.

I.   Recognize the time-management aspect of an individual's team involvement as well as the time of the individual who must cover the job during the team meetings.

FIGURE 11.4 Keeping the team on track.

tions for leaders and facilitators to consider whenever problems occur. Using these illustrations as analysis aids can help leaders and facilitators identify the problems that are occurring.

Team advisors, leaders, and facilitators should periodically get together to discuss team progress, especially early in the team's activity.

Figures 11.7 and 11.8 provide step-by-step action plans for two different family teams over the first several months of team activity. Figure 11.9 provides an illustration of customers identified by an Accounting Department team. Their listing of outputs is far too extensive to share, but its length clearly indicates why the process is never-ending. It's extremely important to reassure teams that they shouldn't be

III. Closing-Out a Team

A.   Prior to preparing their final report, the team facilitator/leader should consult with their advisor for report guidelines, (i.e., content, detail, format, etc.) and to determine whether a formal presentation is desired. At this time an understanding should be reached regarding how recommendations will be considered and when a response regarding their acceptance should be expected. **(The Advisor is responsible for assuring that teams know the status of their recommendations.)**

B.   Indicate who should receive a copy of the written report. Please include a copy for the Quality Process Manager so he/she knows the team has completed its task and can make others aware of the accomplishments.

C.   A meeting for presenting the team's findings and recommendations is important because it provides an opportunity for questions and discussion, as well as an opportunity for recognizing the team for its work.

   1.   Consider inviting others to attend who will be affected by the team's recommendation.

   2.   Please invite Quality Process Manager, if appropriate.

D.   Upon receipt of a team's suggestion, please communicate your review procedure, a completion date for assessing their recommendation, and your appreciation for their efforts.

   1.   It may take several weeks to assess a team's recommendations. Please keep them informed of the status of their recommendations.

E.   Consider informal recognition for a team that has fulfilled its Charter (minutes, milestone chart, team work plan, customer satisfaction process, six steps, presentation/report, etc.)

   Examples:

      - Take team to lunch with Vice President (or appropriate officer)

      - Memo to team members' supervision to be posted in each area complimenting the team's efforts and summarizing accomplishments.

F.   Are there other areas of the organization that might benefit from any of the team's ideas/outputs?

G.   Determine how recommendations should be measured and monitored to identify further opportunities for continual improvement.

FIGURE 11.5 Formal presentation of recommendations and recognition of team's accomplishments.

overwhelmed by the size of their customers/outputs list-ings. Encourage them to select those prioritized as being most important and then proceed with applying the process one customer and/or one output at a time.

ANALYZE THE TEAM CHARTER

(Done first by the Advisor then in discussion
with the Team Leader and Facilitator)

Questions to consider to avoid failure:

_____ Is the charter vague or confusing?

_____ Does the charter define subject boundaries – are there parameters the team should be aware of?

_____ Does the charter include key questions which should/should not be addressed?

_____ Does the team charter identify deliverables? (presentations, written reports, periodic updates, objectives/strategies, recommendations, analyses, etc.)

_____ Does the charter identify target dates for completion?

_____ Does charter provide basis for team to develop a work plan? (set goals, develop action steps)

_____ Does the charter set any expectations for the technical process to be used? (six steps, customer satisfaction process, other)

_____ Does the charter identify the uses of the outputs from the technical process?

_____ Is the scope of the charter too large/small?

_____ Does the charter cover all issues the advisor/charterer feels are relevant?

_____ Can the team work within the boundaries of the charter?

_____ Does the charter convey expectations to the team? (develop team work plan, keep milestone chart, keep minutes)

_____ Does the charter convey what you want the team to accomplish, the business objective to be achieved?

Be sure team activities are not initiated until team objectives and mechanics have been agreed to by the Advisor, Leader and Facilitator.

FIGURE 11.6 Questions for advisors, leaders, and facilitators to consider.

Departmental Quality Implementation Strategy Worksheet*

GOAL: To satisfy customer needs, both internal as well as external, and to continually improve our products, processes, and services.

OBJECTIVE: By (Date), establish department head family teams within the division to determine customer-needs assessment for each department's customers.

STRATEGY: By (Date), establish the (area) family team to assess customer needs.

DELIVERABLE: Present results, orally and in writing, to the vice president.

Action Step	Target Completion Date
1. Prepare charter for the formation of family team consisting of all managers and reports to the director of the department.	6/1
2. Hold first meeting for ground rules and understanding charter.	7/1
3. Determine outputs and customers of the family team.	7/1
4. Attach values to each combination in an effort to determine the most important customer and output of the family.	8/1
5. The most important customer was determined by evaluating the matrix. (Customers were first-line supervisors.)	9/1
6. Identify the customers' needs and requirements. (A brief questionnaire was sent to 35 supervisors asking how we could determine their needs and requirements.)	9/1
7. Result of the survey was to have small focus groups consisting of 5 to 6 supervisors and 1 of the family team members.	10/1

Seven meetings were held with each meeting having three questions asked.

(1) Are communications adequate? If not, how can we improve them?
(2) Do you consider our operations to be reasonably consistent?
(3) If you had one wish, what would it be?

8. Presented all comments for each question.	11/1
9. Condense each group into smaller categories.	1/1
10. Meet with focus groups and review topics within the categories.	2/1
11. Develop strategies to address the topic that had the greatest interest and comment (communications).	2/1
12. Implement strategies that address the topic.	6/1
13. Meet with focus groups to reassess other topics to determine the next greatest topic of interest and evaluate implementation of the prior strategy.	7/1

* Thanks to E.D. Stevenson and Citizens Coke & Gas Utility Distribution Operations for developing this worksheet.

FIGURE 11.7 A step-by-step action plan for a family team implementation.

83

2/22   Formation of team and acceptance of the following mission statement: "To identify and prioritize internal and external customers of the Accounting Department and assess the needs of the priority customers."

2/29   Storyboarded and categorized customer identification (see Figure 11.5).

3/28   Assigned the following categories to outputs:
        Accounts payable
        Administrative and other
        Cashiers
        Financial/statistical reporting
        Payroll
        Property records
        Revenue accounting

7/25   Assessed the outputs and customer identification and prioritized our customers as follows:
        1) Senior management (internal)
        2) Nongeneral accounting areas (internal)
        3) General accounting areas (internal)
        4) Other business customers (external)
        5) Governmental agencies (external)
        6) Utility customers (external)

8/8   Prepared interview format/questionnaire for senior management needs assessment (also prepared a listing of miscellaneous interdepartmental functions for each executive).

8/26   Conducted interviews of the priority customers (some directors were also present) to assess their needs.

10/27   Defined team goals.

3/23   Interviewed the vice president of finance to assess his needs and also provided a needs assessment of the previously interviewed executives.

6/15   Finalized needs assessment to include the vice president of finance's comments. Decided to gather additional information by assessing the needs of other management members by attaching a questionnaire to each report that is distributed internally.

FIGURE 11.8 Milestones in customer needs assessment.

ASSET Team
Customer Classification

INTERNAL CUSTOMERS

100   General Accounting Areas

100	CRDC	105	Payroll Section
102	Credit Union	106	General Ledger Section
103	Intra-Section	107	Property Records
104	Accounts Payable Section	108	Cashiers Section

200   Nongeneral Accounting Areas

201	Public Affairs	206	Financial
202	Information Services	207	Gas Operations
203	Manufacturing Operations	208	Human Resources
204	Gas Marketing	209	Legal
205	Oil Operation		

300   Management

301	Supervisors	305	Managers
302	Directors	306	Vice Presidents
303	President	307	Board of Directors
304	Board of Trustees		

EXTERNAL CUSTOMERS

400   Governmental Agencies

401	Oklahoma Tax Commission	408	Attorney General State of Indiana
402	Indiana Utility Regulatory Commission	409	Regulatory Bodies (Government)
403	Department of Energy	410	Internal Revenue Service
404	Government Agencies	411	Small Claims Courts
405	City Controller	412	Clerks Office Child Support Division
406	State Board of Tax Commissioners		

500   Utility Customers

501	Gas, Coke & Oil Customers	502	Former Gas Customers

600   Other Business Customers

601	Industry Associations	607	Ernst & Whinney
602	Moodys	608	Consultants
603	Standard & Poor	609	Outside Auditors
604	Natural Gas Vendors & Pipelines	610	Gannett Fleming
605	Banks	611	Coopers & Lybrand
606	Vendors (Accounts Payable Customer)	612	Outside Services

FIGURE 11.9 A list of customers developed by a team from the accounting department.

Chapter 12

Form Cross-Functional Teams

Cross-functional teams are comprised of individuals from two or more *uncommon* work areas. Formation usually occurs when management recognizes that the resolution of an issue (problem, need, or opportunity) exceeds the capability of employees from a common work area; consequently, they reach out to include those from other areas affected by or involved in the issue.

Often the need for a cross-functional team will be realized as a family team proceeds through the assessment and into the delivery phases of the quality process.

WORKING TOGETHER WORKS

There is among management and employees alike, an almost instant understanding of the powerful potential of working together across departmental boundaries to resolve issues. Cross-functional team formation may, therefore, occur more rapidly in the beginning for the purpose of solving easily identified problems or needs that are not the direct result of previously applying the process.

Don't discourage this team building, because it will involve employees more quickly than if you wait for the family-team cascade. It also will provide essential first-hand experience applicable to the needs and functioning of any type of team. Also of tremendous benefit to the organization is the breaking down of traditional departmental barriers that previously may have hindered or prohibited employees from different work areas working well together.

A comment for the traditional results-oriented manager: *Cross-functional teams impact positively on the bottom line.* In fact, in the first year of my quality experience, I witnessed one team's annual results pay the entire out-of-pocket costs of our pursuit for three years. Yet, please don't use or pursue the process with a *"savings"* mentality! This can turn off many who can and will contribute *if you keep the focus on the customer.* Be faithful to the quality process and not only will desired results flow but so, too, will come unexpected side benefits.

Cross-functional teams should have advisors, leaders, facilitators, and recorders just as family teams should. Proper chartering of the cross-functional team is even more critical since the family team will be following the quality process steps toward the desired target of customer satisfaction, but a cross-functional team's steps may not be as visible.

Suggestions for forming and closing out cross-functional teams are provided here to share essential guidelines toward successful operation.

I. Forming a Cross-Functional Team
   A. Prepare Team Charter
      1. State who the team advisor or person designated to "coach" the team will be concerning its project, company policies and procedures,

who will offer a recommendation or opinion as a guide to action.

2. Define the subject to be addressed, including key questions which should be answered. Also, include items or areas which should not be addressed.

3. Identify what the team is expected to deliver, including objectives/strategies, recommendations, analyses, etc. Indicate if these "deliverables" will be in report form and/or a presentation, as well as to whom the deliverable will be submitted.

4. Indicate who should be sent meeting minutes.

5. State the expected frequency of periodic update meetings (when the advisor will meet with the team leader and facilitator).

6. Indicate a deadline date for the project's completion and interim steps.

7. Identify what method management will use to respond to the team's recommendations.

B. Select potential team members, designate a leader, and select a facilitator. Be sure supervisors from other affected and/or involved areas are aware of and in agreement with team formation.

C. Meet with the team leader and facilitator to discuss team charter, membership, etc. You might want to attend the first meeting to answer any questions that the team might have. This is very helpful in getting the team off to a good start.

D. Finalize the charter.

E. Send "Team Formation Notification" to the quality coordinator and attach your charter.

II. Closing out a Team

A. Prior to preparing their final report, the team facilitator and/or leader should consult with their advisor for report guidelines (i.e., content, detail, format, etc.) and to determine whether a formal presentation is desired. At this time an understanding should be reached regarding how recommendations will be considered and when a response regarding their acceptance should be expected.

B. Indicate who should receive a copy of their report. Include a copy for the quality coordinator so he or she knows the team has completed their task and can make others aware of the team's accomplishments.

C. A meeting for presenting team findings and recommendations is important because it provides an opportunity for recognizing the team members for their work.

1. Consider inviting others to attend who will be affected by the team's recommendation or who may learn from seeing the methodologies applied.

D. Upon receipt of a team's recommendations, communicate your review procedure, a completion date for assessing their recommendation, and your appreciation for their efforts.

1. It may take several weeks to assess a team's suggestion but keep them informed of the status of their recommendation(s).

The basic charter should be kept simple as illustrated in my chartering memo to our recognition team (see Figure 12.1). As the charter is discussed with the team leader, facilitator, and team members, it should be expanded upon to document the desired understandings established.

Cross-functional teams (or family teams trying to resolve issues common to their work area) must be disciplined in the way recommendations (solutions) are

MEMORANDUM

FROM: Dave Griffiths
TO: Members of the Recognition Team

Re: Recognition Team Charter

Employee involvement is an integral part of the quality process and should include recognizing teams and individuals for their contributions. Recognition should reinforce the behavior that is compatible with the new attitudes represented by the quality process.

The quality implementation team is asking you to:

• Identify the different recognition methods available.

• Determine the appropriateness of each method for our organization, including consideration of the role of the suggestion system.

• Propose a recommended recognition program for the entire company.

• Make your recommendations to the quality implementation team via a presentation and/or a report.

• Make your recommendations by February 1.

• Coordinate activities with other appropriate groups such as the "Communication" and "Implement Quality Process in Manufacturing" teams.

FIGURE 12.1 A memo to members of the recognition team defining team objectives.

formulated. Figure 12.2 illustrates the methodology we have encouraged both types of teams to follow. To further clarify, consider the steps as follows:

The Six Steps to Satisfying Customers
1. Identify the Problem and How It Affects the Customer
   a) Who is the customer? (storyboard all possibilities)
   b) What does the customer want, need, or expect?
      • What needs to be changed?
      • How will the change affect the customer?
   c) What is the problem?

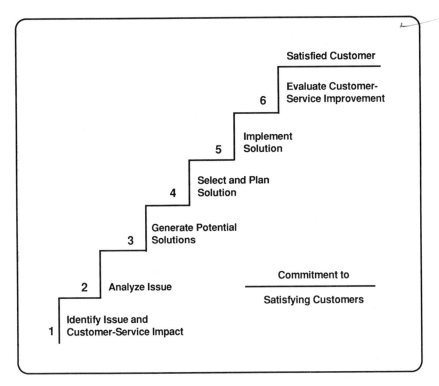

FIGURE 12.2 Six steps to satisfying customers

2. Analyze Issue
   a) What's stopping us from reaching our customer specifications and the desired customer-service level?
   b) What are the key causes restricting desired customer-service level? (brainstorm, document and rank)
3. Generate Potential Solutions
   a) List potential solutions to minimize or eliminate the key causes that are preventing us from reaching our customer specifications.
4. Select and Plan Solution
   a) What solution has the greatest impact on customer service?
   • Is it feasible?

•   How can we measure improvement?
•   How will we implement it?

b) Sometimes the best solution isn't feasible at the time. If this is the case, what can we do in the meantime to bring about change and move toward our customer specifications? (Alternate plan: Things that can be done immediately with minimal effort or cost)

5. Implement Solution

a) Are we implementing the solution according to our plan?

6. Monitor and Evaluate the Improvement to Customer Service—Continue Process *Continuing improvement*

a) How well did the solution work? (Look at measurement criteria)

b) What else can be done to improve customer service? *Sy zero R + P designed to provide continuing improvement of system by ongoing communic. w/ cus*

I have presented these six steps in a customer-satisfaction context because even if the cross-functional team has not been created as a result of customer-needs assessment, it is still extremely important to apply customer-impact consideration. This provides a new element in the prioritizing of issues and is of great assistance in assuring the most important issues to the organization are, in fact, the ones being pursued.

Again, remember that teams are not the end in themselves, neither is employee involvement. The target is total quality . . . to continually improve in satisfying customer needs, and teams provide a major vehicle assisting the pursuit.

Teamwork, however, is important so be sure to convey that as few as two people working together also can be thought of as a team and most certainly can utilize the quality process, isssue-resolution methodology, and the tools and techniques of quality in performing daily tasks.

Chapter 13

Include Individual Involvement

Placing emphasis upon teams or teamwork will cause some employees to feel excluded from the quality process unless they are asked. From the very beginning in your orientation sessions try to convey that each individual has a role in the fulfillment of total quality. When you say "Make quality a routine part of your daily work," make sure employees know you mean make quality the way each of us does our tasks, not something others do or something extra.

GETTING EMPLOYEE BUY IN

As more individuals complete the basic quality education course, expect to hear "Now what do I do?" and "When will I get on a team?"

These are fair questions, and fortunately you don't have to make employees wait. Encourage each individual to use the process and to begin assessing customers' needs.

Have them consider:

- What is my output?
- Who are my customers?
- What are their requirements? (Don't forget to *ask them.)*

It will help if the answers to these questions are written down and discussed with fellow workers. *Save* the results, they can be used at some point later in team involvement.

Next, have the individuals proceed into the other considerations of the process:

- What are the steps in my work process?
- What are the most significant measures of my work (output)?
- Can my work process produce the output desired by my customer?
- How can I improve my output?

Remind the individuals that as they identify problems or issues or opportunities, they should remember to apply the six steps for resolving them.

The best results from the process probably will come when individuals combine their thoughts with others with whom they work and energies become multiplied may times over . . . but the process *begins* with each of us individually considering what we do, who our customers are, and what they need.

A PERSONAL EXAMPLE

A good example of this individual use of the process is my early experience with it. In my assessment phase, I listed my outputs (what I do) and my customers (who I do for).

In thirty years of managerial experience never had I used, or had used upon me, the technique of job-related personal needs assessments. However, after the experience of this first pursuit, I have incorporated this simple but powerful technique into my management/leadership style.

I considered my customer listing and decided that those who report directly to me are the most critical (important).

Previously listed outputs were then analyzed and those applicable to my customers were formed into customer-survey questions, with a one-through-five response scale to permit shading of reactions (see Figure 13.1).

The survey was sent to each of the employees I supervise, ranging from executive secretary to vice presidents. When sent, I had anticipated clustered responses that would identify certain outputs needing attention. I thought I would then conduct a storyboarding session with all present as a family team to assist in developing strategies through which I could better meet their needs. The responses, however, were varied and reflected the fact that each had individual needs requiring specific consideration and independent discussion. (Note: There was, at first, some concern expressed about the idea of "rating" me, but upon reassurance that I was sincere in my desire to assess their needs, there was no further hesitation. Two respondents indicated "a year ago I wouldn't have responded as candidly as now," which indicated to me that the process had had a positive effect upon me, them and/or the work environment.)

Rather than the anticipated joint meeting, individual interviews were subsequently conducted (each respondent had chosen to identify themselves) averaging one and one-half hours in length. During these one-on-one meetings, specific strategies were identified that could better meet their individual needs and/or improve my outputs.

The following lists certain, but not all, needs identified through this assessment activity.

1. Career counseling/planning should be an essential hands-on part of my responsibility. It should not be left up to Human Resources or the employee's own initiative or the assumption that programs satisfy the need.

From: D. N. Griffiths   To: Those I Supervise

RE: "Customer" Survey

As you know, the Administrative Committee is functioning as a family team. We have recently identified our outputs and customers and are now assessing customers' needs. You are my customers, and I would appreciate your responses to the questions below. Each answer should be based upon your feeling as it relates to the work environment and activity for which I am personally responsible. The one-through-five scale (with "5" being your most positive feeling) will hopefully let you indicate whether attention should be given to the topic. Also, your comments and suggestions will be greatly appreciated. Your response is intended to be anonymous but if you want to identify yourself, feel free.

(please circle one)

1. Is your work environment one in which
   you feel you can be successful?   1   2   3   4   5

2. Is our communication with each other
   as often and free as you would like?   1   2   3   4   5

3. Is our organization (administration)
   designed as it should be?   1   2   3   4   5

4. Is quality being promoted in our area?   1   2   3   4   5

5. Are your questions answered promptly
   and satisfactorily?   1   2   3   4   5

6. Is advice given as desired?   1   2   3   4   5

7. Is attention given to your training
   and development?   1   2   3   4   5

8. Are performance appraisals conducted
   as you desire?   1   2   3   4   5

9. Is my leadership as it should be?   1   2   3   4   5

10. When you submit something for my
    approval, do I properly consider it?   1   2   3   4   5

Comments:_____
_____
_____
_____

Many thanks!

FIGURE 13.1 A "customer" survey regarding how well employee needs are being met.

96

2. Developmental initiatives need to be encouraged by the leader (me).
3. Greater structure (i.e., advance scheduling) must be provided to assure time for regular and frequent discussion of activity and questions. (An "open door" isn't enough.)
4. Occasional relaxed time away from the office would complement roles.
5. Performance appraisals should be based upon beginning the evaluation year with mutually developed objectives/targets and mutual evaluation of achievement results, including recognition of "wild cards" (pursuit and accomplishment of job innovations).
6. Organizational structure must be constantly fine-tuned in order not to blunt job opportunity and individual initiative.
7. Communicating as equals enhances roles.

Before using the process I had mistakenly thought that my asking "Now, what can I do to improve?" at the time of the employee's annual review was providing the opportunity for frank discussion, especially since it was only asked after completing the appraisal and salary advisory. The survey, however, provides a substantially better discussion vehicle apart from the appraisal process and a more constructive environment.

In effect, those I supervise and I were working as a team toward how I could better meet their needs. The experience can best be described as being tremendously satisfying, but I believe leading, guiding, and assisting will always be more satisfying than bossing.

Obviously, a manager's use of the process this way sends a powerful message. It undoubtedly reflects a change in the way things are done (unless the manger is truly unique), but it also conveys the manger's commitment to the process and a willingness to explore the unknown. Realistically, the direct report assessment may not be used by very many in the organization, but it does send a message directly and indirectly.

Many other top-down opportunities exist for involving individuals and these will be expanded upon in later chapters. However, I'll touch upon a few to illustrate how an in-tune manager can make the process and its new culture a reality apart from the formal team experience.

More open communication is one such opportunity and with it a broader sharing of information and the abandonment of "need to know" thinking. Managers also can involve many more in fact finding or analysis steps toward strategy formulation. (Remember, be disciplined—use the six steps!) Storyboarding is an extremely useful tool for this activity.

Formal recognition of individuals and their use of the process and/or the satisfying of customer needs provides another avenue toward early involvement, and another great opportunity may be awaiting discovery in your suggestion program. We found ours to be awkwardly restricted by justification procedures, but a team recommended ways to make implementing suggestions considerably easier, which resulted in more being implemented in the first year than had been over the previous thirty years.

Through all of these, try to convey the desire for and concept of continual improvement. All employees can make a significant contribution if they are provided the opportunities to do so.

14

Involve Others

It's extremely difficult to implement quality without out-side support. From the beginning, recognize that your or-ganization will face needs exceeding your internal capability to meet.

CONSULTANTS:
MAKING THEM WORK FOR YOU

Consultants experienced in implementing quality and ap-plying its techniques can bring immediate professional competence to your effort as well as necessary reassurance formed from their past experiences. Over time your needs will change, but be mindful that those brought in from the outside must be made totally conversant with your focus, mission, and objectives. They cannot be permitted to im-part philosophies or methodologies inconsistent with those you've established for the culture you desire.

Be careful not to segment your pursuit by seemingly shifting emphasis away from quality. This will be a real challenge, especially as the effort matures and is cascading to ever-broadening areas. The process must be permitted to take on whatever characteristics are necessary for employees to identify with it, wherever they are located. However, any tailoring that must be done should focus on *how* to best implement the process—not on *what* is to be implemented. This illustrates the continuing need to establish and maintain an organizational *discipline* true to the process, the absence of which will cause a reversion to individualized pursuits . . . to programs and projects.

The role of outside consultants will be to assist you in the development of your quality concepts, educating your employees about them, and then applying them. Make every effort to assure the process is yours, not the consultant's. This will be a challenge, but to effect maximum buy in and commitment, try to achieve a feeling of ownership—a pride of authorship in the endeavor. This will make it much more difficult to walk away from when the going gets tough—as it will from time to time.

Equal to—perhaps even greater than—technical expertise is the need to find assistance from consultants who have the right "chemistry" for your organization. When they communicate, they must be heard. They must be able to turn employees on, not off. They must be able to facilitate not dictate (i.e., they must recognize and meet the needs of their customer . . . you!). They must teach and lead but always with the objective to prepare you to teach and lead. Their intent from the beginning must be to make you self-sufficient, to have you become the expert.

Outside consultants can be an unlimited resource for techniques and technologies. They should be on the cutting edge of developments, yet able to explain them to the neophyte. Many may have forgotten how tedious—and at times how frightening—the early days can be; therefore, they must be able to create a learning environment where the student can *develop,* not be immersed. They must be experienced in quality implementation from its beginning, not just from its application.

INVOLVE YOUR SUPPLIERS

Consultants are not the only outsiders to involve. Vendors of your supplies, materials, and services should be included in your pursuit as soon as you've completed development of your quality concepts and the implementation plan. The following are suggested strategies to assist you:

1. Provide basic quality education to those most critical to your operation. (Later expand this opportunity to everyone.)
2. Reverse roles with them and assess their needs to determine how processes affecting both of you can be improved. (You will probably find ways you can help them be of greater service to you.)
3. Present a quality workshop tailored to vendors and procurement personnel.
4. Form partnerships aimed at continually improving internal and external customer satisfaction.
5. Establish cross-functional teams with major vendors.
6. Encourage vendors to implement quality within their organizations.
7. Routinely audit your vendors, establishing quality standards and expectations beyond the customary quality, price, and delivery standards.

NETWORK WITH OTHERS

Another outside resource to enhance your implementation effort is to network with organizations who are quality practitioners. Those of us who have survived the early days of doubt and struggle with the quality process love to have the inexperienced ask for our wise counsel. We will share war stories for hours, and the most difficult task in the newly formed alliance will be to shut us up, not to get a response. These independent views and experiences can be invaluable not only as a source of how-to suggestions, but also to provide reassurance and examples.

As you grow be prepared to assume a new role in the network as a teacher rather than student, although you'll continue to learn as you reflect upon your progress while assisting others. One of the powerful aspects of quality is its dynamic, continually improving nature. Therefore, even though the basics may remain constant, the way to best apply them is ever-changing. Even as I write, there is trepidation because on the one hand I admit there is much I have learned and can share, yet on the other, I have barely begun to learn.

Other networking agents can be found in and through the formal societies and associations of quality professionals. It seems to me, however, these too often attract the technician, and those of us who are not statisticians are made to feel how ignorant we are. Perhaps some day these organizations can find ways to provide forums for those of us who will never be more than generalists but still need to know better how to make quality happen.

Finally, work toward becoming a leader in the application and practice of quality. Commit to a willingness to educate others at the point in time when you've advanced your effort beyond a gleam in your eye. All of us in this wonderful country must realize we must become more competitive—and quality provides the way.

Chapter 15

Communicate, Communicate, Communicate

COMMUNICATION IS EVERYONE'S RESPONSIBILITY

I have unquestionable credentials to write about communication because I have for some time been responsible for our organization's external and internal communiques. Yet, through my involvement with quality, I have learned more about communication than from all previous experience.

The importance of communication to the process and to the organization cannot be overstated. Most people believe their efforts are already adequate, perhaps even better than adequate. After all, everyone knows the importance of communication. In fact, most large organizations have departments specializing in communications staffed by specialists in the field. This reflects our first mistake.

Communication cannot be left to the Employee Relations or Public Affairs Department or wherever it is functionally assigned. It must be recognized as a task for which we are *all* responsible, not something that is delegated to another area . . . and then forgotten. I am not suggesting you eliminate your communication department or that it isn't conveying information well. I'm simply stating that communication is the responsibility of everyone in management, and we need to be much better prepared in its techniques and considerably more involved in its practice.

Assess Internal Communications

I recommend you form a cross-functional team representing a good mix of your organizational areas, including representation from your communications department. Have them use the process to assess first the internal communication needs of your organization with particular emphasis upon supporting quality. However, don't restrict the charter to only quality needs. In the beginning quality will be viewed as something *outside* the usual activity of the organization, so be sure the team's considerations begin with quality but are not limited to it.

The team will probably convey a major deficiency in information sharing. This should not come as a surprise since most of us (before quality) operate on a need-to-know basis, carefully filtering information as it flows through the company, leaving the grapevine as the only credibly viewed source of information by our employees. Even our communications professionals, when not filtering, are striving to put just the right "spin" on things. Consequently, most of our employees only have minimal awareness of what is actually happening with little, if any, awareness of why.

To illustrate this deficiency, consider the planning activity, which impacts everyone but involves usually only a few in its development with the resulting plan communicated in a restricted manner. Employees are expected to perform their job task without any effort by us to convey its relationship to our corporate objectives and strategies. Yet these are the people who have the most influence upon successful fulfillment and customer satisfaction.

Communication Means Sharing . . . Everything

Now for the pill that may be the hardest to swallow. If you want to implement quality, you must be willing to share . . . to communicate *everything* about your company. Information sharing must be considered as an all-or-nothing issue. If you're willing to share all, you have an important ingredient toward successfully implementing quality. If you're not, I strongly suggest you forget about being a leader and await your competitive demise.

Communication feeds upon itself, and its appetite is never satisfied. The more you do, the more you need to do. The effort is not without reward, however, and communication improvement will be viewed as one of the early positive results of implementing quality.

IMPROVING COMMUNICATIONS

Communications can be improved in a variety of ways, but I'll list a few that I believe to be extremely effective:

1. Make a conscious effort to share all information, being mindful to convey not only *what* is happening but *how* and *why* it is happening.
2. Meet regularly with all employees, not just with those you supervise or peers. (Obviously, in large organizations, CEOs or departmental executives may not be able to have direct contact with everyone, but, to the extent possible, they should try to have some kind of face-to-face information sharing.)

3. Strive to make meetings two-way. Don't simply report, listen, hear and respond. Discuss and share ideas.
4. Incorporate and reinforce the quality mission statement and objectives in every communique.
5. Develop ways to convey quality happenings. (For example: develop a quality "highlights" section in your employee paper; issue regular separate quality reports; display in various locations team pictures, charters, progress reports, results, etc.; use telephone "newslines.")
6. Encourage departmental news memos or papers. (Communiques do not need to be controlled by a single department.)
7. Develop ways for employees to communicate (share) with each other (e.g., forums, workshops, presentations).
8. Use unscripted video commentary from fellow employees to share results, frustrations, and testimonials regarding process worth. (You won't have to manufacture these; they'll be there and not only have a tremendous positive impact upon others but they'll exhilarate you as well.)

As you involve more and more people in the communication function, not only will you greatly expand the organizational intelligence of your people, but you'll develop leaders. There is another benefit, too, in that you make the work environment more alive, more real, more personal, more fun.

Whatever the amount of communication you're doing, do more. Continually strive to improve it. Communicate . . . Communicate . . . Communicate!

Part IV

Supporting the Process

Chapter 16

Recognize,
Recognize,
Recognize

A major component of implementing quality is the recognition of employees who are involved in the effort and have demonstrated noteworthy accomplishment in applying customer-satisfaction concepts and techniques.

Obviously, a recognition program in support of quality acknowledges team and individual contributions, and it also provides additional benefits. Through a recognition program will come role models others can emulate. It also demonstrates quality at work and has the potential to expand buy in greatly, or at least to reduce skepticism.

Certain aspects of recognition are necessary from the very beginning of your effort, but I encourage you to consider using a cross-functional team to recommend the programs and methods best suited to your environment. After all, the recognition need is theirs, not yours, so let them have a role in developing how to best satisfy it. (Employees are internal customers.)

Recognition can and should assume many forms, each of which should be monitored to identify improvement opportunities. (Remember, use the process to further the process.) I believe awards should be part of a recognition program but they need not be extravagant.

TWO TYPES OF RECOGNITION

Even though I recommend you use a cross-functional team to develop your program, there are basically two types of recognition, team recognition and individual recognition.

Team Recognition

Remember, teams are the vehicle for involving large numbers of employees and provide the major thrust for applying the customer-satisfaction process. Therefore it is imperative to recognize teams who have excelled in using the process and who have attained recognizable achievements. This is important not only to team members but it also demonstrates process accomplishments throughout the entire organization.

Steps in recognizing teams are as follows:

1. Solicit nominations once a year from team leaders and facilitators.
2. Develop a selection criteria incorporating the achievement activities you consider more meaningful. (Try to weight "use of process" as heavily as results.)
3. Select teams whose accomplishment will be seen as significant by all employees.
4. Broadcast the news of teams selected and their accomplishments. (Feature them in the company paper, special bulletins, and videos. Team members will love the pub-

licity. Let them tell their success stories, including frustrations.)

5. Use a widely attended recognition event to honor the awardees.
6. Present achievement certificates and awards. (Consider a point system redeemable for gifts or cash since your customers (the employees) will probably be equally divided as to their desired option.) Include something (like a portfolio) that fellow employees will in the future see (i.e., some tangible symbol of honor).
7. Induct honorees into a permanent Quality Hall of Fame which displays team pictures and accomplishments.

Individual Recognition

As the process matures, individuals should accept the process as a routine part of performing their daily tasks. A recognition system can support this concept. Some employees will struggle with acknowledging what is expected in performing their work. I assure you, saying "thank you" and "well done" has never turned anyone off.

MULTIPLE LEVELS OF RECOGNITION

Consider multiple levels of recognition to reflect varying degrees of accomplishment and keep the mechanics simple. I favor an "in the eye of the beholder" approach where any employee can recognize another, but some organizations will prefer award teams or committees. To me this adds unnecessary bureaucracy to what can be an extremely simple, though admittedly subjective, procedure. In any case, the paramount criterion should be to recognize those individuals who have brought a sense of quality and teamwork to their jobs, demonstrating their commitment to satisfying internal or external customers. Every employee in the organization should be eligible, and anyone can initiate the recognition. We use a three-level approach as follows:

Level-I recognition is the most basic acknowledgment of a task or assignment well done. It can best be described as a "thank you" or an "atta boy" or "atta girl" and can be conveyed via a personal note, memo, or preprinted card. This usually is a conveyance between the sender and recipient only, which may seem to dilute the impact but each one I've received has made me feel good, and invariably I'll say "thank you" for the "thank you."

Level-II recognition is more formal and should make others aware of the accomplishment. Certainly, the supervisor of the person to be recognized should be included. This can be accomplished by the initiator asking the supervisor to recognize the applicable individual on his or her behalf. I also think it's a good idea to have a note put in the person's personnel file. In addition to these, use a representative token gift and an invitation to the annual recognition event as ways to elevate this level to a higher plateau.

Level III recognition should acknowledge a truly significant achievement by the individual, usually with meaningful impact upon the organization. The awarding of Level-III recognition should definitely be coordinated with the individual's supervisor. This makes the supervisor more a party to the awarding than to have him or her dictating the selection. However, there may be instances where some eligibility question arises; in that case the supervisor must diligently review the nomination, keeping the initiator informed.

Level-III recipients should have their outstanding performance acknowledged in the appropriate employee news mediums and should receive either cash or an award. They, of course, should have their recognition noted in their personnel file and should be among the most honored guests at the company's recognition event. Induct them, too, into your Quality Hall of Fame.

ADDITIONAL NOTES ON RECOGNITION

There are also some general observations about recognition I'd like to share.

1. Awards do *not* need to be large to be meaningful.
2. The recognition (acknowledgement of achievement) is equal to if not greater than the award.
3. Recognition says someone appreciates someone's effort.
4. Recognition creates role models.
5. Employee-to-employee recognition breaks down departmental barriers, promotes teamwork, and improves communications.
6. Management appreciates being recognized as much as do line employees.
7. A recognition program provides momentum to the process and can keep it going.

Attendance Counts

Finally, I'll share a recommendation made to us by our "TREAT" team (To Recognize Employees And Teams). It was based upon results from their assessment (survey) of employee recognition needs, and it caused considerable debate among our implementation team.

Their recommendation was to include outstanding attendance on a par with team and individual recognition. Management for the most part felt that attendance was more a given than something to be made noteworthy. The team effectively countered, however, that if employees are essential to implementing the process, they must be present to do so, and, most importantly, those who make the extra effort to always be present should not go unnoticed or unappreciated. However, since inconsistencies may exist between hourly and salary time keeping, it may be best to have attendance recognition applicable only to hourly personnel.

Include attendance and acknowledge those with perfect records for one, two, three, four, and five or more years. Have them as honorees at your recognition event, where not only their accomplishment can be shared but where they can hear of the team and individual customer-satisfaction achievements.

Provide the opportunity for the process to build upon itself and for employees to learn by example from one another.

Recognize, recognize, recognize!

Chapter 17

Walk the Talk

PRACTICE WHAT YOU PREACH

I first heard the term "walk the talk" from Ernie Huge, partner in the Manufacturing Excellence Practice of Ernst and Young. The admonishment from Ernie came early in my quality experience and meant less to me then than now. When I first heard it, it sounded cute. Today it is full of meaning; easily said but extremely challenging to practice.

"Walk the talk" is perhaps best translated as "practice what you preach." Many of us, however, lack the discipline to incorporate its intent into our daily habits. Yet to implement quality, we must make our pronouncements real. If executive management can personally demonstrate the quality mission and methodology, our transformation will be instantly seen. If, however, we do not lead by example, and worse yet conduct ourselves in ways that contradict our quality ethic, we will create obstacles of immense proportion.

In retrospect, I believe more time should be spent in the beginning to prepare top management for the new role of leading the organization into a total quality culture. It is not enough to support the CEO's desires or to encourage others to follow the mandate. The CEO's desire or eager expectations can't make quality happen. Management, particularly upper management, cannot be the conduit; they must be the catalyst.

DEMONSTRATE WHAT YOU EXPECT

Opportunities abound for management to demonstrate personally the use of quality techniques, many of which have been described in preceding chapters. I encourage you to apply these how-to techniques to your own work ethic. Permit me to reemphasize a few of them:

1. *Communicate.* Discuss the mission statement with your direct reports, trying to reassure and reinforce your commitment to strive toward making it happen. Together with your employees, consider the quality objectives. Demonstrate an openness to their ideas and concerns. Listen and respond to what you hear. Be gentle and reassuring. Meet regularly. Strive to explain why to what happens in the organization.

2. *Use the process.* Form a family team to assess customer needs and determine how best to meet them and work toward improvement. Use teams to consider opportunities or problems. Incorporate customer consideration into performing daily tasks. Consider how best to implement and support the process in your areas.

   Demonstrate the applicability of the concepts to individual use as well. Include process consideration in your one-on-one sessions with your employees. Determine who your customers are and what your outputs are. Start working to improve your delivery.

3. *Assess the needs of those you supervise.* You should have already identified the employees you supervise as your

priority customers, so ask them how you can tailor your outputs to meet their needs better. In Chapter Thirteen I shared the assessment form I use. It has provided the basis for meaningful one-on-one dialog, and the findings have made me rethink my role and responsibilities.

4. *Become a teacher, a coach.* Let go of what you think is authoritatively and rightfully yours. The years you've spent in getting ahead in the organization have been filled with learning and experiences that you should share. Start developing those who will lead your organization in the future by sharing authority and responsibility. Challenge yourself to delegate as much as possible, then advise and guide the new generation of decision makers. Frame discussions with the employees you supervise around quality concepts and techniques. When you begin to receive questions about the process, it will be viewed as being part of everyone's "regular" job.

   Encourage innovation and, in particular, continual improvement in processes and in individual performance.

5. *Recognize those achieving customer satisfaction.* Every day look for opportunities to say "Well done!" or "Thank you." A note or a simple spoken acknowledgment can have an extremely positive impact.

   Use your newly developed recognition program to convey to everyone the good deeds of those around you. Strive to create an environment of positive reinforcement.

   Promote and reward those excelling in quality.

6. *Get away from the desk.* Don't continue to be victimized by an inbasket schedule. Make time to be out and about. You can't "walk the talk" sitting down. You'll be surprised at how much you'll learn about what's going on.

   Managers who "walk quality" are directly involved in the process. They understand it and use it. More importantly, conveying the quality philosophy, spirit, and feeling is every manager's responsibility.

Chapter 18

Monitor Process Progress

The organization must maintain an ongoing awareness of its progress in implementing quality. Pursuit strategies should be documented and routinely reviewed by those directly responsible for coordinating your effort and the quality implementation team.

Beyond meeting stated implementation action steps is the more significant need of identifying improvement opportunities. Some of these will surface as problems that must be addressed, others will come as suggestions and recommendations from quality innovators within the organization. These innovations can be facilitated by encouraging an awareness of what other organizations are doing through seminars, articles, books, etc.

USE BALDRIGE AS A GUIDE

A more absolute framework, however, is required to assure a common direction and to minimize, if not eliminate, debate over priorities. The Malcolm Baldrige National Quality Award criteria, as previously described in Chapter Three, provides an important consensus of quality professionals that can be effectively used for assessment of your achievements and needs. The guidelines are rigorous but they establish standards essential for successful total quality implementation.

Using the Baldrige criteria, your organization should conduct annual audits of its quality progress. These should be of two types: external assessments conducted by knowledgeable professionals from outside your organization, and internal assessments, conducted by a team consisting of your organization's employees. The combined perspectives will provide a meaningful view of your progress and, more importantly, recommendations for continued growth and improvement. To appreciate fully the detail and the depth of the criteria, you should obtain and study the full publication.* In it you will find a description of the issues to consider and examine in seven broad components. My intent here, therefore, is not to summarize an exceptionally well-presented guide. I do, however, want to share my thoughts regarding these criteria.

Leadership

An active, participating role by top management is essential to implement quality effectively and achieve long-term improvement. Executives must demonstrate their commitment to and involvement in the process and the new quality philosophy.

Employees naturally will be skeptical about management's buy in or conversion to a new way of thinking and doing; therefore, "walking the talk" is vital. Hints on how

---

* The *Malcolm Baldrige National Quality Award Application Guidelines* are available from the National Institute of Standards and Technology, Route 270 and Quince Orchard Road, Administration Building, Room A537, Gaithersburg, MD 20899.

to do this have already been provided in the previous chapter, but there are actions that will add credence and support to those already described.

- Support teams using clear charters, establishing progress checkpoints, and maintaining momentum.
- Communicate throughout the organization both short- and long-term visions of your quality pursuit and progress.
- Make quality presentations to groups outside your organization, and write articles about your experience.
- Become a student of quality techniques.
- Ensure that your organization's quality training addresses both team and individual needs.

Information and Analysis

The information-and-analysis category examines the way data are used to formulate plans and guide the company. Using the criteria in this category also aids in determining management's "statistical thinking" capability.

Although all organizations have well-intended operational measurements in place, too often they are not customer-oriented and may not reveal customer satisfaction. Customer-satisfaction data often are considered on a departmental basis rather than in a cross-functional, organization-wide manner.

As you analyze your operation in the framework of this category, you may find existing data are founded in productivity consideration with little, if any, linkage to true end-customer satisfaction. You may, therefore, be measuring the wrong things.

In addition to striving for a greater customer orientation in your data, also:

- identify critical customer-based success factors
- involve many more employees in measurement activities
- strive to develop data to support pro-active rather than reactive actions
- communicate any trends to the rest of the organization
- provide measurement training, including its presentation to the entire organization
- develop measurements to determine the cost of non-quality (nonconformance)
- involve suppliers in developing quality and service data

Strategic Quality Planning

The Baldrige guidelines describe this category as the criteria for examining "the company's planning process for retaining or achieving quality leadership and how the company integrates quality-improvement planning into the overall business planning."

I believe this statement gives the impression that quality and quality planning are issues separate from daily operations. Strategic quality planning isn't something to be integrated into business planning; it should drive long-term, short-term, departmental, and personal planning. I feel a customer focus rather than an improvement focus causes this to happen naturally.

Therefore, as you assess your operations planning activities, consider them in the context of the degree to which they are targeted toward continually improving the satisfying of customer needs.

Specific considerations should include the following:

- Vision, mission statements, goals, and objectives should be focused on satisfying customer needs. (Remember, not

only do you have external customers but internal customers as well. For example, your internal and external
customers include employees, stockholders, and regula-
tory agencies.)

- Customer satisfaction must be considered when setting
  goals, objectives, and strategies in order to assure proper
  prioritization for decision making related to improvement, projects, and budgeting.
- All employees should be educated in planning techniques and involved in developing plans.
- Goals and objectives should not be developed using conventional wisdom. They should be based on an assessment of customer needs.
- Both setting of targets and measurements must be based
  upon customer needs assessment.
- All planning mechanisms should be linked to one another so that at every level—from corporate to individual—everyone is able to understand why the action is
  being pursued.
- The planning process should be recognized as a delivery
  mechanism to satisfy specific needs.

Human Resource Utilization

The Baldrige guidelines state that this "category examines
the effectiveness of the company's efforts to develop and
utilize the full potential of the work force for quality and to
maintain an environment conducive to full participation,
continuous improvement, and personal organizational
growth."

It is important to recognize that human resources issues
are not subjects to be addressed only by human resource
professionals. Management must realize that the development of an environment conducive to full employee in-

volvement and total quality is a responsibility shared by all, and can be facilitated by:

- maximizing the involvement of all employees in the process.
- educating all employees in quality concepts and techniques on a continuous basis.
- establishing programs to recognize employees for their contributions to quality.
- considering employees as customers, assessing their personal and environmental needs, determining how best to provide the systems to meet these needs, and then measuring and monitoring achievements to develop improvement strategies.
- considering whether existing human resource functions such as appraisals, salary administration and training are a complement to or are in conflict with quality.

People are an organization's most important resource, and as such, systems must be designed to support and enhance their involvement. Employee input must be more than encouraged and considered, it must become an integral part of the decision-making process.

Quality Assurance of Products and Services

The Baldrige guideline's explanation of this category states that it "examines the systematic approaches used for total quality control of goods and services based primarily upon process design and control, including control of procured materials, parts, and services." It also examines "the integration of quality control with continuous quality improvement."

In many ways this area will be viewed in traditional

"quality control" terms, however, here again I point to the importance of customer focus.

After assessing your customers' actual needs, don't be surprised to find some specifications, standards, or procedures that are not what they should be. For example, we learned that the consistency of our product was more important than several of the characteristics under careful quality control.

The Baldrige criteria recognize the importance of customers by addressing how customer needs and expectations are converted "to product, process, and service specifications." When this is the first consideration, other aspects of quality assurance are framed properly.

Since this is a typical area, I assume all organizations have probably more than enough methodology to address it; therefore I'll not dwell upon the obvious. What is essential, however, is to:

- make sure your measurements and standards are established by assessing customer needs.
- develop quality-audit methods consistent with your new total quality thinking.
- recognize quality assurance applies to all outputs not just to those your traditional customers see.
- establish a broad base of appreciation for the importance of satisfying customer needs *before* measurement techniques are introduced.
- be pro-active not reactive in your quality assurance pursuits.
- be dissatisfied with simply meeting standards. Always look for ways to improve continually. If you don't find them, your competition may.

Quality Results

This sixth element of the Baldrige criteria points to the fact that quality doesn't "just happen," and emphasizes that we must know how we are doing, with attention upon more than just the bottom line.

The results looked for here are of measures "derived from customer needs and expectations."

Trends are of particular importance to your quality pursuit. Even though achieving a total quality environment may take five to seven years, early in your pursuit you should:

- identify key product, service; and process measures.
- update these key measurements each year.
- revise your key measurements as more is learned about actual customer needs.
- compare the measurements, particularly those that are process-related to those of other organizations, especially those considered to be leaders in their field. (Don't think you can't find them, you can!)
- establish this benchmark as one of the ways to identify improvement opportunities and develop applicable strategies. (The improvement strategies are important, not the comparative numbers.)

Customer Satisfaction

This is the seventh and final category of the Baldrige criteria. Within these areas are a total of twenty-seven subcategories for which achievement points are awarded, totaling one thousand points. "Customer Satisfaction" elements comprise thirty percent of the total awarded. (Interestingly, "Quality Results" and "Human Resource Utilization" account for another 30 percent. Therefore, these three are the most significant in the minds of our country's quality experts.)

Through the "Customer Satisfaction" element we are asked to "examine the company's knowledge of the customer" and how we meet customer requirements and expectations. In all, there are fifty-six "areas to address" described within this element contained under the subcategories of:

- "Knowledge of Customer Requirements and Expectations"
- "Customer-Relationship Management"
- "Customer-Satisfaction Methods of Measurement and Results"

Allow me to emphasize once more the potential of using the Malcolm Baldrige National Quality Award guidelines to frame your quality pursuit and measure its progress. The fifty-six customer-satisfaction subcategories identify issues that will more than challenge your organization and its system.

It is no wonder so few have been awarded the prize, but the true award is in the pursuit.

Part V

Enhancing the Process

Chapter 19

Convert Management to Leaders

I should have noted at the very beginning of this book to "Read Chapter Nineteen first." Often I'm asked, "What is the single biggest challenge or problem in implementing quality?" (Often this is followed with a second question, "Is it the union?")

The answer is that it is we who manage who present the greatest challenges to the implementation of quality. I know I was never taught in school or during thirty-plus years of on-the-job experience what I've come to know these last three years. The biggest challenge in implementing quality is that of converting management . . . *especially upper management* . . . to leaders.

CREATING A LEADERSHIP ENVIRONMENT

Look at the definitions of "manage" and "lead," and the contrast will become instantly apparent. Although some of the same words are found in each, the tones are considerably different . . . and so, too, are the management styles. Unfortunately, most of us have developed styles that are "to direct" and "to control" rather than "to guide" or "to show the way." The fact is we need to spend much less time bossing and much more time teaching. Consider what you're doing in performing your daily managerial tasks, and ponder whether you're paying attention to the truly important or to minutiae. Too many of us have become slaves to our desks, telephones, and meetings. When we are called upon to direct the needs of quality, we respond, if not openly at least inwardly, that "we don't have time." The truth is, for the future of our companies, there is nothing more important than creating an environment where quality can flourish . . . and that requires *leadership.*

As part of your quality education, therefore, establish early in your pursuit coursework for all supervisory personnel in leadership skills and techniques. Much of what is required is behavioral as depicted in Figure 19.1; however, there are specific actions that can be instantly incorporated to move you toward transformation.

Form your own family team—Applying the quality process and techniques in a team setting with you as the leader (coach/teacher) will move you instantly toward the desired environment. Identifying your customers and outputs and striving to meet customer needs will break down barriers and create cross-functional attitudes.

Conduct a personal-leadership assessment—Have your subordinates and associates evaluate your leadership skills. From the evaluation, identify improvement opportunities and develop applicable strategies.

Get out from behind the desk—Analyze the work that has you deskbound and eliminate or delegate wherever possible. Use the extra time to get out and about. Observe operations . . . not just your own but of the entire or-

Leadership Characteristics

Coach/Teacher	Walks the talk
Leads by example	Practices open communication
Good listener	Stresses honesty
Has integrity	Is self-disciplined
Appreciates human values	Willing to try new ideas
Flexible	Delegates
Team-oriented	Trusts

- Has a clear understanding of the purpose and direction of the organization and shares same.

- Commits totally to satisfying internal and external customer needs.

- Recognizes the efforts of others.

FIGURE 19.1 Leadership characteristics, skills, and techniques.

ganization. Talk with your fellow employees. Find out what's going on. Hear what they have to say.

Use your corporate and departmental plans as teaching vehicles—Discuss with your employees the goals, objectives, and strategies of the formalized plans. Help them understand "why" things are done.

Establish individual targets linked to the plans—Mutually develop personal strategies early, complementing or supporting the formal plans that can be monitored on a regular basis (no longer than quarterly).

Incorporate continual improvement into task expectations—Add "continual improvement" to job descriptions and convey an expectation of task accomplishment that includes improvement expectations.

Communicate regularly—Formally and informally share information with employees about what's happening in the organization.

Meet direct reports on neutral ground—Schedule regular breakfast or luncheon meetings (ideally off premises) to chat about how things are going.

Become directly involved in developing your people—Find out the career desires and ambitions of those you supervise. Discuss ways they can augment their potential; counsel and guide them.

LEADERSHIP STYLE

As you work toward becoming a leader, anticipate certain issues of style to have a greater impact than others. Among these I would expect to find the following:

Issue: OPENNESS

- Are you candid with the information you share?
- Do you listen to the input received?
- How well do you work with other managers/departments?
- Do you encourage conflicting viewpoints?
- Do you avoid risk?

Issue: TRUST

- Do you treat employees as you would like to be treated?
- Do you recognize employees as customers?
- Can problems be brought to you without fear?
- Are your appraisals/promotions "fair"?
- Do you support your employees and their ideas?

Issue: DECISION MAKING

- Do you encourage decisions to be made at the lowest levels?
- Do you truly delegate?
- Do you encourage teamwork?
- Do you encourage data analysis in decision making?
- Do you respond to recommendations?
- How quickly do you respond?

A greatly improved leadership style—like quality—doesn't happen overnight so we must commit to a never-ending pursuit. We must always be learning about and striving to incorporate the necessary ingredients. Without rigorous attention, we can easily lose what was so difficult to gain. Leadership skills must, therefore, be kept finely honed and always in use. Otherwise, like muscle, they become soft, flabby, inactive.

For pure enjoyment of daily work, nothing can compare with leading rather than bossing. Leading "makes you feel good," and its rewards will be many moments of euphoria.

Recognize that management will need to be taught leadership skills, so don't leave conversion to chance. Again, "educate, educate, educate" must be accepted as the way to make the necessary changes occur. Utilize outside professionals to assist you in this endeavor and accept the fact that you'll need a series of workshops, seminars, and classes to break old management styles and habits.

Although leadership is an individual-by-individual challenge, approach the conversion as a management team. The dynamics of team learning and sharing are as valuable here as they are elsewhere. As an executive family team, we identified our employees as one of our major customer groups. From assessing their needs (remember, use the process), we identified specific improvement opportunities—many of which were directly tied to leadership or the lack of leadership. Following are the types of issues you can expect to be raised:

- Lack of delegation/empowerment
- Lack of trust
- Inadequate/bad appraisals
- Poor communication
- Poor listening
- Limited personal growth/organizational opportunities
- Career counseling
- Inconsistency with quality pursuit

Addressing these issues as a team takes you instantly into the major processes and systems of your organization. Use the executive team to consider the ramifications of the leadership characteristics listed in Figure 19.1. These are certainly topics professional training should address but also make them discussion topics for all levels of management. This means that other managerial teams should also be involved in developing the corporate understanding of management's role and responsibility in implementing quality.

Much is said about the changes in corporate culture needed to implement quality, but remember—in order to achieve these cultural changes, management behavior must change.

Chapter 20

Develop and Use Meaningful Techniques & Measures

In Chapter Nine I discussed basic techniques essential to implementing quality. The purpose of a second chapter seemingly about the same topic is to separate those things you must do early from those things that you'll grow into.

QUALITY ESSENTIALS

Just as you accept the process as ongoing, so, too, must you recognize that quality concepts are developed over time as is recognizing what is—and isn't—meaningful in the support of quality. Allow me to outline the process techniques I believe to be essential to achieving successful implementation.

Previous chapters have covered some of these topics, but I want to provide a single snapshot of the most important quality ingredients.

Establish a Customer Focus

You must have an absolute customer focus recognizable by and acceptable to *all* employees (from executive through line). I am unrelenting in my conviction that for universal buy-in, the focus must be singular in nature, not of multiple translation. By this I mean "quality improvement," "problem solving," or "productivity improvement"; all have different meanings, depending upon who hears the term and/or considers its impact.

The *customer focus* and, more specifically, satisfying internal and external customer needs is the only one able to stand on its own merit (while at the same time incorporating the varying desires and objectives of all organizational levels).

Give Equal Attention to Satisfying Internal Customers

Instant opportunities abound if attention is given to internal customers and the outputs and processes related to satisfying their needs.

Every organization is shackled by inadequacies in interdepartmental relations. Turfdoms will at first erode and eventually disappear where quality with an internal customer focus flourishes.

In addition, many existing individual and departmental outputs will be questioned as to their value and efficiency in procedure and flow. For example, when internal teams look at customer needs related to information, an impressive number of reports will be eliminated or reduced.

Establish a Methodology

Quality doesn't just happen by willing it done. You must have an understandable process (Chapter Two) that can be systematically followed. Through its use, all can strive to become more disciplined.

Educate Everyone on an Ongoing Basis

Various chapters discuss education, and it's essential to recognize we all must be better prepared in order to implement quality. Everyone must understand the basic concepts and then be shown how they are applied. Team leaders and facilitators must be trained, and management must be unrelentingly shown how to lead, support, and use the process. With quality our education never ends . . . and what an excitement this brings to the workplace. The excitement of always learning how to improve continually.

Use Teams as the Single Most Powerful Implementation Vehicle

Total quality cannot be implemented by a few well-intentioned folks. It requires broad-based involvement. Teams provide the structure where assessment, analysis, and understanding of needs and opportunities will flourish.

Develop Individuals to Support the Process

Teams need leaders, facilitators, recorders, and advisors trained in support techniques such as statistical process control (SPC), quality function deployment (QFD), benchmarking, and data gathering. Identify those who excel in technical and support techniques, and use these individuals as internal consultants.

This process infrastructural "need" has significant organizational opportunity in it as you provide an environment where individuals can grow and mature beyond traditional career-ladder/job-advancement aspects.

Incorporate Quality into your Planning Process

Many will view quality as "something extra we've been asked to do," so the challenge is to bring it into our everyday routine. I believe the planning process provides this capability and, moreover, that quality should eventually drive planning at every level.

Customer-satisfaction improvement objectives and strategies should be what our plans are all about.

Establish a Formalized Implementation-Assessment Strategy

You must have an implementation plan against which you can monitor progress. Internal and external assessments using the Malcolm Baldrige criteria will provide annual evaluations of your progress. From these will come the objectives and strategies for implementation refinement.

DATA GATHERING AND ANALYSIS

The more commonly used quality-assurance techniques must support the broad implementation techniques. During the time I spent learning about and implementing quality, I often wondered how someone, myself included, with no statistical aptitude could so suddenly develop such an appreciation of meaningful data.

I think the answer, at least for me, is that statistical techniques seemed too complicated, weren't needed, and didn't fit with the requirements of successfully accomplishing my job tasks.

Assuming there may be others like me who are reading this (hopefully)—and for those of you who are the statistical quality professionals—let me offer some suggestions as to how data gathering and analysis can be made more relevant to the quality process.

Naturally, people must be taught how to use the techniques that support and facilitate the quality process. Too often, however, people are taught *how* to use the technique without enough attention to *when* and *where* to use it.

The formalized quality process methodology (assess, deliver, and monitor) focusing upon customer satisfaction and utilizing the development of strategies for continual improvement provides the opportunity to maximize the usefulness of traditional quality assurance techniques.

Figures 20.1, 20.2, and 20.3 add this final "techniques and tools" dimension to the previously depicted "activity" and "questions to consider" hints shared earlier. These illustrations provide a complete guide to the quality process to which I so often refer. (Please remember I define quality as "satisfying internal and external customer needs.")

Linking tools and techniques to an established methodology so they are seen as one is only part of our challenge. We cannot expect our "students"—our fellow employees—to grasp instantly through a short course the importance of data to decision making. Also, most will be rather timid about using these techniques in real situations.

We must consequently reinforce the initial exposure with hands-on guidance and coaching (remember this is part of adding leadership to managing). Also, look for or make opportunities for other employees to describe their use of techniques in arriving at recommendations. Beyond this, be sure to include consideration of tools and techniques in future quality refresher courses. Demonstration and education become, therefore, ever-continuing.

Finally, there is the issue of meaningful measures and their use. Early in your pursuit have all departments select a few of what they consider to be their most significant indicators of operating performance. Gather these at one location and use them as base points for future reference to analyze process progress.

Customer Satisfaction Process Guide - Assess Needs -

ACTIVITY	QUESTIONS TO CONSIDER	TECHNIQUES/TOOLS TO USE*
Identify Customers, Outputs List them all – there is no right or wrong answer!	Who are the customers? What do we provide?	Storyboarding
Select Customer(s) or Output(s) to Focus Upon If you are just starting pick one – save the tough ones for later.	What criteria should be used? Who are the key customers? What are the key outputs? Do you have different types of customers? Have you discussed the results of this step with your team advisor?	Pareto analysis Voting with dots What/How matrix Charts
Identify Customer Requirements Start with what you think is the customers' need, then ask them.	Do you really know what your customers want? Are there other needs that customers have not stated? Have you negotiated the requirements with your customers? Have you reviewed current and past data? Do your customers all have the same needs?	Surveys Focus Groups Face-to-face meetings Industry data Interviews Internal consultants Suggestion box
Compare Requirements (Whats) and Outputs (Hows) Organize the data and do some analysis.	Are there needs with no outputs? Outputs with no needs? Are the specifications or measure for the requirements? How do your customers feel you are doing? How do your customers rate you against the competition? How do you feel you are doing?	Six steps What/How matrix Benchmarking

*See "Memory Jogger" Booklet

FIGURE 20.1 Techniques and tools for assessing needs. (Developed in conjunction with Ernst & Young.)

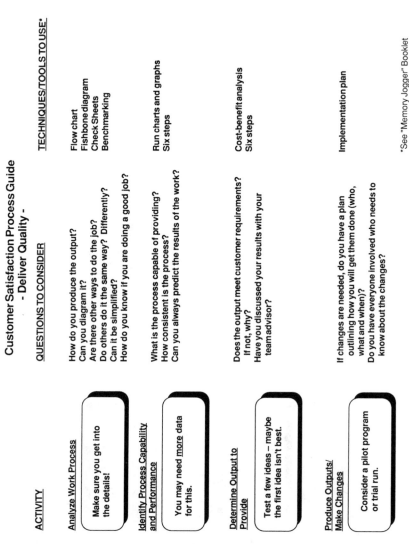

Customer Satisfaction Process Guide
- Deliver Quality -

ACTIVITY	QUESTIONS TO CONSIDER	TECHNIQUES/TOOLS TO USE*
Analyze Work Process Make sure you get into the details!	How do you produce the output? Can you diagram it? Are there other ways to do the job? Do others do it the same way? Differently? Can it be simplified? How do you know if you are doing a good job?	Flow chart Fishbone diagram Check Sheets Benchmarking
Identify Process Capability and Performance You may need more data for this.	What is the process capable of providing? How consistent is the process? Can you always predict the results of the work?	Run charts and graphs Six steps
Determine Output to Provide Test a few ideas – maybe the first idea isn't best.	Does the output meet customer requirements? If not, why? Have you discussed your results with your team advisor?	Cost-benefit analysis Six steps
Produce Outputs/ Make Changes Consider a pilot program or trial run.	If changes are needed, do you have a plan outlining how you will get them done (who, what and when)? Do you have everyone involved who needs to know about the changes?	Implementation plan

*See "Memory Jogger" Booklet

FIGURE 20.2 Techniques and tools for delivering quality. (Developed in conjunction with Ernst & Young.)

Customer Satisfaction Process Guide - Monitor Results/Continually Improve -

ACTIVITY	QUESTIONS TO CONSIDER	TECHNIQUES/TOOLS TO USE*
Monitor Process/Outputs Identify Key Success Factors Get enough data to give a clear picture.	How do you measure performance? Where is improvement needed? What does the team advisor think? What should you monitor to guarantee that your product or service will continue to meet the needs of the customer?	Charts, graphs Benchmarking material Measures Fishbone diagram
Identify Competitors or Best-in-Class Brainstorm possibilities then narrow down.	Who is the best in marketing, financial, cost, quality, and delivery?	Research, interviews Storyboarding
Gather and Analyze Data Get into more detail – reexamine cause/effect relationships.	Can you improve the process or output? What have you learned from your competitors or others with similar processes?	Benchmarking material
Develop Improvement Strategy Plan your work before working your plan!	What is your improvement target? When and how do you plan to do it? What help or resources will you need? What are the action steps and timetable? Are you continuing to monitor the process?	Six steps Strategy work sheet
Continue the Process What do you want to work on next?	What is the next priority? Should you go into more detail on the same issue? Should you pick other processes or outputs?	Pareto analysis (see assess needs) Results of research and interviews

*See "Memory Jogger" Booklet

FIGURE 20.3 Techniques and tools for monitoring results and continually improving. (Developed in conjunction with Ernst & Young.)

Some three years or so into your effort you may be surprised to find certain of these no longer have the significance once thought and that other more meaningful measures have been identified. This transformation is a natural consequence of the shift in your organizational focus to customer needs.

Too often we fail to make the most of the data available to us, so be sure your organization considers carefully the measures being used to assure that data are meaningful in a customer-focused quality environment. Once again, the Malcolm Baldrige Award guidelines provide proper direction for how we should be making the most of data as analytical tools for identifying where improvement opportunities lie.

We need to strive constantly to get *inside* our numbers. Here again analyzing data over several years often provides illumination of occurrences sometimes missed by year-to-year snapshots. Data can provide pictures of what is happening, and if we develop a constantly probing mind-set . . . always wanting to know why . . . our plan objectives and strategies can take on a new meaning.

Chapter 21

Link Personal Performance to Quality

From the beginning one of the major challenges to success-fully implementing quality will be that of assuring execu-tive buy in.

TIE PAY TO PERFORMANCE

This can't be left to chance or individual determination. Don't trust either in gentle persuasion . . . no matter how artful it may be. The intent to implement must be absolute, and there is nothing more absolute for the purpose of conveying the intent than to tie pursuit and implementation to the paycheck.

Some will argue against a tied-to-pay approach, but be-lieve me, it is an effective way to convey implementation expectations and degree of importance. Organizational commitment to the process will be more clearly seen when those in the top two levels of management understand that they have a personal stake in assuring that more than lip service is given to the pursuit.

Establish measurable activities important to quality implementation, then "negotiate" short- and long-term personal-accomplishment targets. Always be mindful to stress the importance of the process and its techniques and recognize the numbers relating to exposure don't indicate degree of actual utilization and acceptance. Still, there must be top-level understanding of executive responsibilities. To accurately describe these, specific objectives must, therefore, be established, using such activities as:

- employees educated within a certain time
- quality teams established
- employees involved in team activity
- numbers of customers/outputs identified and addressed
- establishment of customer-based measurements
- benchmarking pursuits
- improvement recommendations developed
- leadership effort being exerted

Incentive systems must reflect the importance of implementing quality so that rewards exist for those who pursue and achieve it. This has another advantage, too, because it makes quality a part of the routine, not something "extra."

You will note my emphasis throughout this chapter has been upon quality support at the executive level. This is because the upper levels bear the highest responsibility for implementation success. Upper management must provide the work environment superstructure where quality can flourish. If we provide the foundation and framework, our employees can—and will—do the rest. If we don't, our area(s) will be the worse for it.

Quality is not pursued without risk, and although the steps toward achieving it are simple, there must be a discipline to its pursuit. This is another reason for supporting

the linking of personal performance to successful implementation, because it is something upon which a few of us
have enormous influence.

DON'T LET RISK SCARE YOU

Please don't use the risk factor as a rationale for not pursuing quality. Risk can be greatly minimized, or eliminated, if we pay attention to certain details, one of the most
important of which is true executive involvement. Having
used the frightening word "risk," let me quickly admonish
that the greatest risk related to quality is in not pursuing it
because those who don't will be less able to respond to
change, to improve, or to be competitive. The organization
that doesn't pursue quality will in time find it has been sur-
passed—and will probably wonder why.

MAKING TIME FOR QUALITY

There is another element supporting linking executive personal performance with quality implementation. Each of us
demonstrates organizational priorities by the emphasis we
place upon daily tasks. Quality cannot be translated as
something "extra," it must become the routine. One of the
major hindrances, however, to quality becoming a routine
part of doing business is in the early stages finding—or
making—time for quality.

Since from the beginning everyone involved will question how time can possibly be found for education, training,
team meetings, added communicative briefings, one-on-one
discussions, walking around, etc., I hesitate to admit to the
fact that implementing quality requires hours upon hours
of involvement. The time commitment cannot be ignored,
so face it boldly but with confidence. Time can—and will—
be found if executive management desires to find it.

The truth of the matter is that whether you analyze the work habits of management or of the entire organization, there is a great deal of wasted time spent doing the unimportant. Recognize the existence of time inefficiencies and convert the problem into the opportunity to do more within the eight hours available than ever before dreamed. Quality also can bring fun and enjoyment to the workplace and with it added energy to be applied to that previously thought of as being routine.

How can all this be done? Management should collectively and individually analyze daily work, questioning how and why their routine is done. Among the inefficient activities to be found will be:

· authorizations required at too high a level
· meetings without agenda or resolution
· discussions without focus or purpose
· in-baskets with trivial pursuits
· numerous unneeded paper handling
· antiquated telephone and message systems
· paper shuffling
· filing
· unneeded reporting

Many of these impact others in the organization so the time wasted is compounded, often many times. One of these of major consequence is report preparation and distribution. Do you want to hazard a guess about how much time your organization spends daily in preparing, copying, distributing, and filing unnecessary reports?

The point in all this is to acknowledge that quality does take time, but it doesn't need to be time not already available. Management can lead the way by initiative, analysis, and change. Months and months later you'll be astounded when you reflect upon how many hours have been spent that somehow came from somewhere *while operations dramatically improved.*

Here then is another example of where leadership is required, . . . where "walking the talk" must be demonstrated. We tell ourselves we are "managers," yet often we fail to manage effectively what is most important. We focus on decision making, problem resolution, and the bottom line but fail to recognize the impact we can have upon these if we pay attention to the working environment, our organizational infrastructure, our employees.

These issues cannot be manipulated or delegated. They are topics for which we at the top of our organizations are responsible, so why shouldn't our personal performance be measured—and rewarded—against our record in their achievement?

We argue, perhaps, that results from these are not measurable, but such simply is not true. Specific targets can be established; employee attitudes and needs can be determined; time commitments can be recognized. While doing these, analyze the traditional bottom lines whether they are profit, efficiency, scrap, accidents, or absenteeism . . . and be amazed at what happens.

Executive management's performance should be first and foremost linked to how we pursue and implement quality.

Chapter 22

Establish the Concept of Continual Improvement

CONTINUAL IMPROVEMENT IS EVERYONE'S RESPONSIBILITY

When stated it sounds so simple, . . . so commonplace. "Establish throughout your organization, within each department and within each individual, the understanding of and appreciation for continual improvement."

Look at your job description, then look at the job descriptions of those you supervise. Is there any mention of improving the tasks described? Chances are your answer is "No!" because traditionally we have defined the job to be done with statements of the specific activity but have not included recognition that we are to be constantly pursuing ways to improve upon the work requirements.

Consequently, productivity improvement is left to be translated as something to be feared . . . as something removed from your daily responsibility . . . as something to be searched for and identified by outside specialists or consultants . . . as separate projects or programs initiated by management at the expense of others' job security.

It shouldn't be this way. Continual improvement should be viewed as the responsibility of each of us in the fulfill-ment of our daily work. The search for it requires a dili-gence—a commitment—but it also, I believe, requires a methodology having a focus easy to identify and easier still with which to identify. The concept of and methodology for identifying customer needs and continually improving in the way the needs are satisfied as described in this book pro-vide the ways to go about this.

More is required, however, than the emphasis first con-veyed in your basic quality education and training. Re-member, individuals must grow in their understanding and use of quality. The concept of continuous improvement is not, therefore, something introduced early, but instead is best conveyed perhaps a year or two after the general un-derstanding of assessing customer needs, ensuring the de-livery system provides the needs, and measuring and monitoring the degree of satisfaction.

IDENTIFY OPPORTUNITIES FOR IMPROVEMENT

Assess, deliver, and monitor are the three phases of the quality pursuit methodology, and it is in Phase III, "moni-tor and measure," where we seriously apply the rationale for measurements . . . to identify opportunities where im-provement strategies should be developed.

There is a great tendency to establish the "improve-ment" from the beginning, but I argue this is not something easily bought into by many in the organization because of the job threats implied and "it's broke and must be fixed" translations.

The full power of continual improvement can be real-ized most when it is viewed in the customer-satisfaction context. So be patient and let the process develop to the point where the thinking of continual improvement has an "acceptable" place and is the next logical step in the transition to a total-quality environment.

Early in the third year of the quality effort another total work force educational experience can provide reinforcement and clarification of the basic concepts and methodology of quality. This next course also can move the organization toward the more sophisticated, mature application of the process to continuously strive to improve customer satisfaction by individuals as well as departments and the organization.

Ernst and Young, in assisting Citizens Gas in the development of its QE3 (third course in quality education), helped us in our continual improvement understanding by describing the attributes of continuous improvement as compared to those of innovation (see Figure 22.1).

Innovation is, of course, the mind-set developed for us as we are taught to think of improvement. Traditionally, we are a "most bang for the buck" industrial society rather than one of "slow and steady wins the race."

Innovation – Find the big improvement idea with the big payoff
Focus is on cost/benefit (is this worth the investment?)
Study-and-analysis oriented
Most often done by specialists
Looking for the "big bang"

Continuous – Make small improvements to every aspect of the
Improvement business on a continuous basis
Focus on no-cost / low-cost ideas
Action oriented – Do it, test it, try it . . .
Quick implementation
Everyone participates

FIGURE 22.1 Innovation vs. continuous-improvement strategies. (Developed in conjunction with Ernst & Young.)

CREATE AN INFORMATION-RICH
ENVIRONMENT

There is another important reason for waiting until your organization grows into a broader understanding and acceptance of quality. Continuous improvement flourishes best in an information-rich environment where information about the business is widely available and freely exchanged inside the organization . . . where turfdoms don't exist. It requires an environment where ideas are welcome and acknowledged, rather than seen as faulting or threatening. The continuous-improvement environment makes the most of teamwork, and management encourages cross-functional sharing of information.

These attributes are seldom present in an organization in the beginning, therefore, they must be nurtured over time. This requires the leadership of management because only we have the power to make or break the concept.

Paraphrasing from the Ernst and Young/Citizens Gas QE3 text:

Management must:

- be supportive with an emphasis upon improving the capabilities of employees and systems toward satisfying customer needs.
- desire and pursue an integrative operational approach where communications and ideas flow freely through the organization.
- place a strong emphasis upon teaching and coaching.
- encourage collaborative, not competitive, work styles.
- support and recognize both individual and group contributions.
- reinforce disciplined use of the quality process
- accept and encourage continuous education and training.

Only management can assure the organizational infrastructure where problems are seen as improvement opportunities. Also, it is only management that can assure improvement pursuits are framed within correct time parameters. Continuous improvement is not a race but, like grains of sand in an hourglass, over time its results will be seen.

Hopefully, you now have a sense of what to do but probably still wonder "How?" as far as the mechanics of continuous-improvement implementation are concerned. Fervent pursuit of quality will cause many of the requirements to take place but specific concentrations will facilitate your efforts.

- Assess customer needs regularly.
- Measure your delivery systems and processes against the best to be found.
- Analyze and improve (use a cross-functional team) your suggestion system.
- Implement recommendations quickly.

Too many organizations hinder improvement opportunities with burdensome "justification" requirements requiring multiple bureaucratic approvals. Rethinking the justification issue can result in a concept of "do it unless there's a reason not to."

In a one-year pilot program designed by one of our cross-functional, multi-level teams, the following occurred. (Please note the pilot was limited to only *half* of our total employee population.)

- Almost as many suggestions were made in one year as had been made by *all* employees over *thirty-four* years!
- More suggestions were implemented than had been in the previous thirty-four years!
- Fifty-six percent of all eligible employees submitted improvement recommendations

- The average number of recommendations submitted per participating employee was seven!

Although few of these were of major cost-saving consequence, the overall impact has been remarkable, and the transition to broad-based improvement thinking is an outstanding accomplishment.

Karl Albrecht and Ron Zemke in their book *Service America!* state: "The ability to understand the customer's needs and wants can be summed up in a simple phrase: 'Always be learning.' " To this I add, . . . "and always be improving."

Part VI

Process Power

Chapter 23

Positive Results and Benefits

Many may be skeptical of the positive potential of a total quality pursuit. Others will see it as being too risky to tinker with an entire organizational structure even though they may pride themselves on being risk takers. It was not uncommon to find quality initiation in the early '80s driven more from desperation than by design.

I'm reminded here of one of my early childhood readings called *Keystone Kids* about the change to a last-place baseball team caused by a fantastic shortstop/second base combination. In it is this great statement by the team manager, "When you're sleeping on the floor, you can't fall out of bed!"

The point is you don't have to be "sleeping on the floor"—to be desperate—to begin the journey. The enlightened executive will recognize the competitive advantage to be gained through implementing total quality. Yet there may remain a hesitancy, perhaps because of the total commitment required or because of a project or program mentality.

This chapter is for "ye of little faith" or of "faint heart." In it I will share some of the happenings over a three-year

157

span at Citizens Gas & Coke. I do this with pride on the one hand, reluctance on the other. You see, I've become a purist regarding quality in that my conviction today is that if an organization concentrates upon implementing the quality process (with a customer focus), the positive results will naturally happen. I, therefore, encourage you to be true to the process and to let the entire organization know of your faith in quality as a process rather than maintaining a traditional results orientation.

"INTANGIBLES"

Results, however, there will be, both tangible and intangible, the latter of which I think may be far more long lasting and of greater potential, so I will begin with these but not in any order of significance.

Increased Awareness of Customer Needs

The organization that is customer-driven will prosper. Further commentary is found throughout this writing. However, for supporting evidence I point to hard evidence used later.

Increased Employee Involvement

Teams and a "new look" to considering recommendations and suggestions change the way people feel about the workplace. In three years we have educated everyone in the basics. Thirty percent have taken the additional "Leaders/ Facilitators" course, and we're now conducting a third total organization education in applying quality concepts. In addition, forty-five percent of all employees are or have been involved in the team activity, and we anticipate the fourth year to increase significantly the involvement.

During our third year, fifty-six percent of all eligible employees submitted improvement ideas to the team-revised BTU (Brilliant Thoughts Unlimited) program, averaging more than seven suggestions each. But of greater significance, more suggestions were implemented in one year than in all previous thirty-four years combined.

Seeing Other Departments as Internal Customers

Functional work areas, thinking of other departments or the next area in the process flow as being their "customer," provide results often difficult to measure directly but of substantial consequence upon organizational measures. The broad customer focus, which includes the internal concept, leaves no improvement stone unturned.

Improved and Enhanced Communications

Organization-wide information sharing provides the database upon which improvement ideas are built by teams and individuals.

The quality organization will have a culture and work environment where people understand the corporate mission and how they relate to it . . . where people, regardless of title, talk with each other and listen . . . where a common goal is sought and the information for its pursuit is available.

Employee to Employee Recognition of Effort

Recognition provides a momentum all its own to the quality pursuit and with it a compounding of accomplishments.

Each of us acknowledges that our juices flow whenever we are praised for a task well done, but, too often, we forget to fuel further the next efforts of others with positive reinforcement of that just done.

Expanded and Enhanced Organization-Wide Planning

A major, positive result of the total quality pursuit is its impact upon the entire planning activity.

Many times more people will be directly involved in plan development but of equal significance is the substance of the planning effort and the potential for linkage from the highest level of corporate goals down through the organization to short-term individual targets.

In substance, the sharpening of design focused upon improving satisfying customer needs through assessment and measurement provide the direction toward competitive dominance.

"TANGIBLES"

For those who demand "hard" numbers, I urge you to study the success stories of the Baldrige Award winners. These leaders of U.S. industry should be viewed as the role models of our quality efforts. In each case significant benchmarks will be found that can be used as achievement targets.

In most cases the Baldrige Award winners' experience comes from six to eight years of intensive effort, so, at best, my personal "results" experience is only half of theirs. Allow me, therefore, to use the accomplishments of two quality "masters" as my first example of what quality with a customer focus can mean. Both are 1989 Baldrige Award winners and have impressive credentials supporting their selection.

Xerox Corporation

Xerox has been our quality mentor from the beginning because of their customer focus, which includes internal as well as external customer-needs assessment and improvement effort.

Xerox results to date* (I'm sure they would be the first to tell you that winning the Baldrige Award is not the end of their quality journey) are as follows:

- More than seven thousand quality-improvement teams involving approximately seventy-five percent of all employees
- A seventy-eight percent decrease in machine defects
- A forty percent decrease in unscheduled maintenance
- A twenty-seven percent reduction in service-response time
- Significant increases in customer satisfaction (Xerox is the leader in five of six satisfaction categories)
- A major reduction in scrap and per-unit costs

Milliken & Company

Although I'm not personally acquainted with the Milliken quality process, it, too, has provided noteworthy results for the South Carolina-based textile manufacturer.

- A seventy-seven percent decrease in supervisor-to-labor ratio (all employees are referred to as "associates" at Milliken)
- A sixty percent reduction in the cost of nonconformance
- An increase of on-time deliveries from seventy-five percent in 1984 to ninety-nine percent in 1988

---

*Source: "1989 Award Winner Xerox & Milliken," *Business America*, November, 1989, pp. 2–11.

- Forty-one quality awards from their customers over five years, including several from U.S. firms noted for their in-depth quality audits

As you examine the Baldrige awardees, note that in every case their expectations for further improvement are more dramatic than their past achievement. These are companies who are ever-honing their competitive edge.

One of the marvelous aspects to the utilization of customer-focused quality concepts is that their application is not limited by organization size or type. Large or small . . . for profit or not-for-profit . . . corporation or association . . . manufacturing or service orientation . . . private or public, the improvement opportunities are present.

Although Citizens Gas & Coke is just into the fourth year of its implementation effort, the results are worth noting.

- One team's annual savings exceeded the total out-of-pocket costs of the quality pursuit.
- Customer satisfaction has improved from fifty-nine percent favorable to seventy-six percent favorable.
- Satisfaction measurements compared to those of other utilities now exceeded the utility average in thirteen of fifteen categories and eight of these with at least a ten-percentage point spread.
- Accident incident rates have been significantly reduced.
- Grievances and arbitrations have been dramatically reduced.
- Customer audits of the manufacturing operation have improved at an annual rate of six percent with two at ninety-eight percent and one hundred percent the past year (two of which were the highest ever awarded).
- A one-year pilot of a new suggestion program designed by a quality team resulted in more recommendations implemented than the total for the previous thirty-four

years. (Fifty-six percent of all employees eligible participated.)

- Cross-functional teams have successfully addressed a wide range of issues, including oil-production improvement, uniform redesign, shift rotation, in-house contracting, and education needs.

In my mind, the major benefits of implementing quality are yet to be realized. The ability for all employees to know and understand our missions, to be educated in techniques supporting its achievement, and, in particular, to be involved—to have the opportunity to work toward its fulfillment—promises even greater results for the future.

Chapter 24

Challenges

Implementing quality is not an easy pursuit, and it's important to anticipate the challenges you will encounter along the way.

In the beginning you should expect expressions of doubt regarding the need for or the organization's ability to commit or respond to the demands of quality. There will be those who genuinely believe they are already doing improvement, teamwork, involvement, or communicative activities to the degree possible or needed. Others won't believe there is a competitive or strategic need for a long-term quality commitment. A related aspect to this expression is the recognition by many in top management of the personal commitment required viewed in the context of their remaining years before retiring (i.e., "Having reached the top, why should I create anew? . . . devote such time? . . . run the risk?").

NO TIME IS NO EXCUSE

There will be feelings or concerns of inadequacy from some who would "like to, but. . . ." These spoken, or more frequently not spoken, thoughts often are based in very personal frustrations stemming from actual and perceived job demands. "Time," or, more correctly described, the "lack of time" becomes in these instances the instantly grasped rationale for not accepting or even trying quality. We who have done "it" are viewed, it seems, as probably having had time on our hands to begin with . . . or not having demands in our jobs equal to those of the beholder.

Let me reemphasize that time is a major challenge in the beginning and will remain so until quality becomes the routine way business is done. Said another way, quality will at first be the unnatural way, done outside the norm, and the challenge is to make it the normal way issues and opportunities are addressed. It may help to offer, as part of your leadership instruction, coursework in time analysis and time management because time will need to be found, but it is there to find for those who know where and how to look.

QUALITY IS PART OF MANAGING

There is another issue compounding the "no time" expression. Many of us see our jobs in too narrow a view and don't understand the breadth of our management responsibility. We, therefore, see the necessary direct and indirect supportive demands of quality as being foreign to what we are supposed to be about in our decision making (i.e., "bossing") roles.

Too many times, I fear, we ascend the organizational ladder as though it is a pyramid with ever-decreasing responsibilities in those activities initially accepted when we were first-line supervisors. We think our higher responsibility is now of policy and that others, usually entire departments of others, will address the new demands required to support quality. Therefore, we argue that we "don't have time" to involve others, or to analyze, or to communicate (listen), or to teach and coach. Ponder this point for a moment and think about how impersonal we have allowed executive task fulfillment to become and about how we've failed to recognize that we must still have a "hands-on" style as we advance . . . but in a different way. To illustrate this point further consider how much or how little time we spend with the employees we supervise in goal setting, progress review, information sharing, performance appraisal, and career counseling. These aren't the only people-related tasks that will most likely be found to be lacking, but they're probably the most glaring examples.

What is an even sadder indictment is that those people in our organization whom we describe as being "exceptional" or "distinguished" routinely are involved in doing the all-important people-related activities. We recognize this on the one hand yet don't realize the extent of our responsibility to those who are not people connected. The challenge, therefore, is to teach managers how to be more involved with their people and what they do; how to advise, guide, and support; and how *not* to become slaves to their offices, telephones, and corporate meetings.

QUALITY IS EVOLUTIONARY

Another time challenge is the issue of "pace." We are not conditioned to pursue with a "slow and steady" mind-set. We expect results right now. Yet, you must recognize and work to accept the fact that quality implementation must be evolutionary. Its pace is not sudden or rapid. By its nature it must be done a step at a time, always assessing how well your delivery system is meeting the needs of your internal customers . . . always fine tuning, looking for ways to improve its delivery. We never know all there is to know; we must always be trying to learn more. This "never reaching the end"/"never being finished" is challenging at best for many, nearly impossible for others.

Time will challenge in still another vein, this one being that of momentum. Through the early days of implementation, perhaps for the first two years, the pursuit will have a momentum of its own. Since the process will flow from the top downward, it also will have, in the beginning, the characteristic of being "close" (i.e., physically near) to those responsible for implementation. About the third year this begins to change and implementation assumes a momentum of its own. During this time expect to have feelings of uncertainty about "how it's going" and/or "what's happening out there." You'll no longer be directly involved in every facet of the quality process and will need to devise methods to measure and monitor (remember, the third phase of the quality process) progress and happenings.

QUALITY FROM THE TOP DOWN

Tracking team growth and individual involvement by area will be of assistance, but tapping into team activity and status may be a challenge in itself. One technique used by our CEO at our weekly staff meeting is to have each of us first report quality happenings in our respective areas. This, of course, requires us to request updates from those we supervise and so the request flows down through our organization. This, obviously, sends the message of interest from the top to all in supervision.

(By the way, teams themselves will cause a furrowed brow or two because even the best-led team will have moments of members' frustration and some teams will even fail in their mission. You must be on your toes to support, rescue, or bury, whichever may be the appropriate course, but in every case guard against the negatives experienced becoming "it's not going to work" rationales at which those who doubt or criticize can point.)

Management commitment—or the assuring of it—may be the biggest challenge of them all. Earlier in the book I stressed that "we," not others, are the hurdles and sometimes the obstacles that stand in the way of implementing quality. Therefore, assuring management's personal involvement and commitment from the beginning and beyond is a responsibility of the organization's or unit's chief executive. There must be developed a *will* to achieve quality . . . a desire to make it happen.

The way to do this is not difficult. Those who have this responsibility must be demonstrative in their commitment. At every opportunity talk quality, its mission, objectives, and results. Remember, though, you must do more than "talk" the concepts, you must use them (Ernie Huge's "walk the talk"). Make those who report to you "think" quality by establishing and conveying your expectations of their responsibilities in making it happen. Put them on the hook and keep them there.

Make the hook sharp. For executive management have at least one personal-performance objective impacting upon pay that is directly related to quality (e.g., team formations, people involvement, recommendations implemented, results attained). If you have organization-wide incentives, establish "implement quality" as one of the major elements to be pursued, and reward its achievement appropriately.

There are also "softer" ways to support and reinforce quality implementation and practice, but they are nonetheless effective. Use the power of recognition to the fullest. "Attaboy" and "attagirl" notes of appreciation for individual effort have significant impact as do other "strokes" previously described. Make opportunities to do some of the recognition publicly—before fellow employees. (Our CEO recently implemented "The President's Club" to single out those doing quality above and beyond. He always makes the presentation before employee groups and never on an expected basis.)

Promotion is another reinforcement tool. See to it that those who get ahead are those who are quality practitioners, and be sure it is known that this is the way to get ahead. (There is an obvious danger or challenge within the promotion opportunity because if the one promoted is not concerned about quality, the message sent throughout the organization will be loud and clear.)

Through all of these challenges is one underpinning element and that is "discipline." When so much of quality is common sense, why is it that its practice is so abused? I believe it is because we are terribly undisciplined in the way we conduct ourselves—at work and otherwise. The customer-satisfaction process I've described works wherever there is a desire to make it work. Unbelievable results can be attained wherever there is the discipline to follow and support the process.

In the work setting we are more fortunate in our ability to force discipline; we can bring to bear much more than our pleadings or rationales. The reward-and-recognition opportunities previously described coupled with an individual's desire to achieve can provide the catalyst needed provided we keep the focus of our pursuit recognizable to all. The quality mission statement, objectives, and process must embrace all that we do, and we must constantly guard against inconsistencies. Too often our work processes are seen as segmented, non-related activities. Quality can pull everything together, providing a common cause, a common methodology, a common language, and a common appreciation for our achievements.

Chapter 25

Opportunities

When I first outlined this book I had challenges and opportunities as a single chapter, for indeed every challenge is, in fact, an opportunity. Each of the issues described should be addressed in the context of the six-step issue-resolution methodology (Figure 4.4) beginning with analysis of the issue and its customer impact (remember, customers are also internal), followed by development and eventually implementation of improvement strategies.

Beyond the potential of successful resolution of the identified issues are opportunities of enormous consequence toward assuring long-lasting competitive advantage. We who lead, after all, have a strategic responsibility to our organization, and even though our attention seems to be constantly toward focusing upon the immediate, usually in the form of the bottom line, the long-term health and vitality of the company is very much in our hands.

THE NEXT GENERATION

Quality is an investment for the future. An investment that fortunately pays immediate dividends, but one where the largest payoffs will be experienced later. (I have estimated it takes five to eight years for full implementation to occur; however, Xerox, in accepting the 1989 Baldrige Award, stated it is a ten-year effort.) Admittedly, therefore, writing of future benefits is conjecture on my part, but I am totally confident in my forecast.

By now you should recognize that total quality as I have described it is a complete change in organizational culture. As the change is occurring—while quality is being implemented—the organization is preparing for and developing toward a far more dramatic and responsive state, the major elements of which are its people and the way they think, feel, and act about job tasks.

Throughout this company of the next generation there will be a spirit of participation and involvement. Satisfying customer needs will be universally practiced. Use of the quality process and its techniques will be routine. Of major impact will be the fact that all of your corporate leaders will have a different management style than the one we were first taught.

Strategically, this new generation of leadership is one of the more dramatic results of implementing quality, and, fortunately, you won't find yourself waiting long to enjoy it. Within a very short time frame, the education of all exempt employees in techniques of leading and facilitating teams coupled with the hands-on experience of team involvement will generate a talent reservoir awaiting opportunities for greater responsibility.

Complementing this occurrence will be the changes taking place in your planning activity, leading eventually to a linkage between corporate, departmental, and individual goals; budgets clearly driven by quality improvement

priority; and organization-wide awareness of and involvement in developing strategies for continual improvement.
It is here, in the pursuit of continual improvement, that your
organization will find the greatest opportunity for "implementing quality . . . with a customer focus."

When my editors first critiqued the early chapters of this
book, they felt greater emphasis should be placed upon
adding value for the customer. There should be no question
in your mind regarding the priority of "adding value" because the truth is that those who go beyond satisfying basic
customer needs will be the corporate success stories of tomorrow.

CUSTOMER-SATISFACTION FLOW CHART

In Chapter Four, I described the customer-satisfaction process. Figure 25.1 provides a less-detailed depiction of the
process than that of Figure 4.1 but for our purposes here
the simplified version works best. As you consider the flow
chart, recognize that the actions related to assuring that the
work process satisfies customer needs, the use of the six-
step methodology for issue resolution, and finally the search
for and development of improvement opportunities are all
steps toward adding value to our outputs.

In Chapter Twenty-Two I describe at length establishing
the concept of continual improvment. Those first steps are
the basic "how to" directions toward adding value, which
in the beginning will be in the context of meeting customer
needs, but having the potential to eventually take you beyond what is expected . . . to exceed and consequently excel.

Summarizing earlier statements, I believe there is a risk
or at the very least the potential for not realizing the full
benefit in focusing too early upon value-added issues. Those
who approach quality focusing immediately upon work-
process analysis, problem solving, comparative measurement improvement, or even continual improvement are doing

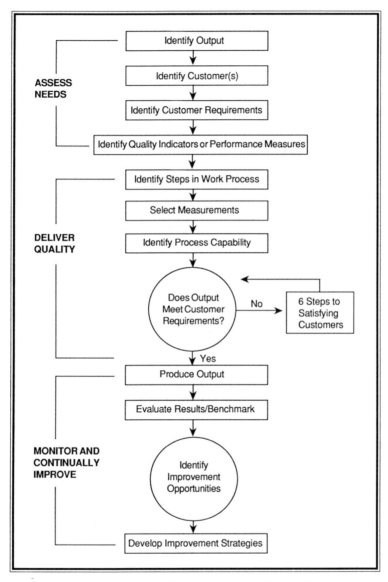

FIGURE 25.1 Customer-satisfaction process flow chart.

173

so too early. When there is general acceptance of the importance of knowing customer needs, the aforementioned pursuits can occur. As you look, therefore, at Figure 25.1, realize
there is a sequence to the events described that is indeed a
flow. There is a relationship in each activity to those preceding it, and to embark upon one of the later elements
without addressing the earlier stages presents the strong
likelihood of achievements not being of the substance or
duration possible.

Here, again, the subject of discipline is at hand, and I
challenge you to maximize the opportunities that are yours
by nurturing the implementation of quality so that it and
your organization mature together and realize all the benefits that can be had.

Perhaps another way of arguing my point is to say it is
indeed the third phase, "monitor and continually improve,"
that we want to do. It is our destination, but we must travel
some distance before we get there. So much for my personal philosophy, now let's concentrate upon Phase III, accepting my argument to do it at the appropriate time, and,
to please my editors (for they are quite correct) let's add
substance to the "monitor and continually improve" description and instead call it "add value."

NEVER ENDING CHALLENGE

Constantly adding value will be, for those who pursue it,
the competitive edge. There are two key words in this statement: "constantly" and "pursue." Said another way, "continually improve." Inherent to the expression is
acknowledgment that "it" is *never* achieved. We never stop
trying to add value; we are never satisfied or content with
how well we're doing. We accept as a given that improvement is ever-attainable, and that as the current provider we
hold the competitive advantage to be the probable improvement innovator.

Herein, therefore, is the first aspect of how to add value beyond the obvious improvement opportunities. You must have a mind-set, a determination to search forever to find a better way. As a leader you must provide a work environment where this can occur. You must challenge those within the organization to stretch beyond the way things are done today. You must inspire and you must reward not only those who achieve but perhaps, more importantly, those who try but don't. It is the *pursuit* that is first and foremost because with pursuit will come accomplishment.

Beyond this, constantly measure and set benchmarks. Know not only what your best has been but how it was achieved. This means getting inside the measurements so you are not addressing what is outside the limits in a negative way, but are trying to learn more about the positive variances from the norm.

Since you've come into the improvement mode *in sequence*, I accept as a given that your measurements are "correct," that they are the significant, critical success factors. I've already discussed benchmarking, but it may be that you'll need to develop benchmarking partners and with them refine your measurements. This should not be viewed as collusion but should be seen as competition in its finest form where measurements of success, significance, and satisfaction are agreed upon. All partners are then challenged to improve upon them—not to the public's detriment but to its benefit.

In the first application of continual improvement I urged you to downplay innovation; at this stage you should now seek it through research and development. You also will have discovered, provided you are observant, that some of the employees who have been provided the opportunity to be involved have excelled in their contributions. They have, in effect, become innovators. Make use of these employees as a valuable resource, enhancing your ability to be more than common, and, more importantly, to assure you stay ahead of your competition. To some, all of this may be more clearly understood if viewed as though we have added a fourth phase to those previously described. I prefer, however, to view it as a continuum—always in motion and never ending.

The adding of value to the dimensions described is, therefore, the ultimate opportunity the pursuit of quality provides. It is far beyond where we are today. Attainment is not something "perhaps possible," it is something definite. For as certain as I am of the benefits to be gained through quality, so am I certain that what we do today will be done better next year, and ten years from now it will be done in a manner not thought possible today.

Chapter **26**

Continuing the Pursuit

You've now been presented a step-by-step guide to "implementing quality . . . with a customer focus." Each part of this book is in essence a "phase," and each chapter denotes a major activity to be accomplished. These are not sequential and overlaps do occur. Figure 26.1 depicts the implementation phases in a time-line approach, reflecting when related steps began or are anticipated in our approach. At this writing we're completing our fourth year of implementation with the needs for the next two fairly well defined.

MAINTAINING MOMENTUM

The purpose of this chapter is to offer suggestions on how to keep the pursuit going. Another way of looking at this is in the context of how to maintain momentum. Using Figure 26.1 as a pictorial guide, I'll therefore discuss pursuit/momentum based upon what I've experienced and expect.

During Phase I there are relatively few in the organizaton directly involved; consequently, the pursuit or effort is

Phase I:
Investigation, Selection & Development of Quality Concepts (Focus, Mission, Objectives, Definitions, Process Elements, Implementation Plan)

Phase II:
Basic Education and Preparation (Employee Orientation, QE1 & QE2)

Phase III:
Applying the Process (Team Formation, First Use of Process, Demonstrating Individual Applicability)

Phase IV:
Support for and Enrichment to the Process
(Leadership Development, Communication, Recognition Personal Performance Objectives, Internal Consultants, Internal/External Assessments)

Phase V:
Enhancement to and Sophistication of Process
(Continual Improvement, Performance Measures, Benchmarking, Linkage of Quality Improvement to All Plans)

Phase VI:
(Commitment to Satisfying Customers is Routine Part of Daily Work and Improvement Exceeds Expectations)

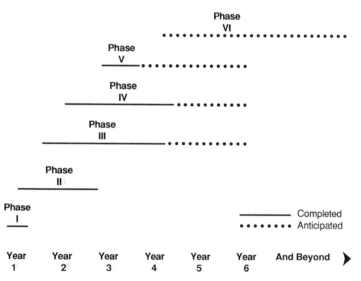

FIGURE 26.1 Development of customer-satisfaction commitment.

a somewhat routine task not at all unlike typical organizational projects. Although it is an extremely important period, it is, as indicated, very short in duration and the happenings within it define themselves.

Phase II is likewise straightforward and has as its beginning the announcement of the effort, followed by educating all employees in basic quality concepts (QE1), and training those who will facilitate and lead teams (QE2). The time required to do this is directly proportional to the resources applied as related to your total employee base. This time line could be measured in hours for a very small organization or months for a larger organization. The issue of *momentum* will first surface during this time because a feeling of anticipation will be created and within it will be a mix of hopeful expectations, uncertainty, trepidation, and skepticism. The completion of the tasks of this period will outpace your response capability, so, in effect, a "pull" is created that will have you straining to support the desire for involvement.

Phase III, therefore, overlaps with the activities of Phase II. As soon as the first employees have completed QE1 and the first leaders and facilitators have been trained in QE2, team formation should begin. It is here a pinch point occurs in what I'm sure has been my easy to follow "how to" guide. I've simplistically described the use of the process with customer-needs assessment as its first step in a family team (common functional work area) applicational mode. A conflict in my methodology arises because all employees can't be educated or team support personnel prepared at once. Further, there is a tremendous advantage in "mixing" class participants . . . management with rank and file. Consequently, even though I urge a top-down rollout of the process, the most realistic approach in the beginning is to let some education occur for a slight time *ahead* of initial team formation. This pause will also give you time to fine tune the training classes, incorporating the assessment recommendations of the participants (course "customers").

Even though you try to "manage" the formation timing of your first teams, don't do it to the degree of discouraging those who can see the potential in a team addressing an important issue (problem or opportunity). If you concentrate upon assuring teams are properly chartered and supported (advisor, leader, and facilitator) and that they are disciplined in the use of the quality process, the first teams being focused upon cross-functional issues will present no problem.

As quickly as you can, however, form family teams. Do this by requiring your first two upper-managment levels be mixed into your first QE1 and QE2 classes. As soon as a majority of them have "graduated," the CEO should form his or her family team, followed shortly thereafter by team members forming their own teams and so the cascade (rollout) should begin. As family team implementation moves further into the organization, always be sure a majority of team members have previously had the basic quality education course (QE1) and that those who have not are quickly scheduled into future classes.

During Phase III, implementation will acquire a momentum of its own during which your challenge will be to assure the pace is supportable. Remember, "support" includes your ability to respond to team recommendations, and key to this is management's readiness to receive new ideas in a noncustomary way. For this reason I encourage you to *immediately* follow QE1 and QE2 with coursework that I'll describe as Leadership I for *all* management personnel (plus I'd include all other exempt employees as well). This, therefore, is the beginning of Phase IV during which your attention will be upon those actions supporting and/or enriching process implementation.

Phase IV also is particularly important in our "how to continue the pursuit" consideration because it contains many of the issues that, if properly addressed, will assist greatly in providing ongoing momentum. For this reason, permit me to reemphasize certain points already covered in previous chapters, critical not only to support and enhancement, but also later on to your ability to continue.

Leadership

You'll find a few "naturals" in your organization, but for the rest of us, specialized training is required to present concepts and techniques in guiding, advising, coaching, teamwork, communications, and especially listening. A new set of attitudes must be developed and a willingness to involve and/or delegate established.

Communication

You can't do enough communication and it is not something for which others are responsible. Beyond keeping the process in focus, various communication techniques are essential to convey happenings and results while always reminding of the quality mission. Information and dialog are essential in the quality organization. Consequently, improving communications is something all management must pursue.

Recognition

Devise ways to recognize teams and individuals who excel in the use of the process and who clearly demonstrate the achievement of customer satisfaction. As does "communication," the development of recognition methodology is an excellent charter for an early cross-functional team. In such an effort management can put the process to use, involve many never previously involved beyond customary job tasks, and perhaps most importantly can *implement* team recommendations (a form of recognition in itself).

Assessment

Formal assessments or audits of implementation progress also are critical to your success because they provide the identification of improvement needs and opportunities. You should naturally monitor your status against a previously documented implementation plan, but approximately a year to a year and a half into your pursuit, and at similar increments thereafter, closely examine how you are doing, using the Malcolm Baldrige Award guidelines.

These four issues (leadership, communication, recognition, and assessment) deserve your attention early but should be recognized also as vital components through which your present effort can be supported and future continuation enhanced. Upper management personal performance objectives directly related to implementation and support and impacting upon salary and bonuses, establish the organizational implementation intent and rewards for accomplishment. These, too, should be initiated early and continued. Finally, within the second and third year, certain individuals will be demonstrating quality aptitudes and capabilities that should be utilized. Identify and make use of these various talented internal consultants, recognizing that they not only provide needed team and individual support, but many will be the future leaders of your organization.

Phase V is a period where quality implementation begins to lose its separate identity. On the one hand this is excellent because it means quality is nearing the point of being routine. From the implementor's view, however, Phase V presents new challenges because the pursuit is approaching being everything that is done in the company or organization. "Momentum," therefore, will have a great tendency toward moving in different directions, and great care and attention must be given to keeping the course steady.

Fortunately, the steps taken in the earlier phases provide much of what is needed. Don't, however, consider these previous taks completed. Revisit them time and again, always using the process to improve the process. They become, therefore, the foundation and the ongoing support elements for continuance. In particular, I want to stress again the value of the Baldrige criteria. Judging your progress compared to it and identifying improvement opportunities do, in fact, provide the capability for further strategy development.

In Phase V, the pursuit clearly becomes focused upon improvement, but in the context of customer needs. Your planning process will become the major vehicle for translating improvement targets into meaningful objectives and strategies. Measurements will then be in use that have been developed through better understanding of what is needed to satisfy the customer and through your efforts to benchmark with others who may have found better ways. Your analysis capability will become much more sophisticated and in all likelihood more straightforward as your measurements are refined.

Clearly, one of the most significant aspects of reaching this point in your process development is the change that will occur in your plan. Your strategic plan should by this time be visible to all but, more importantly, a linkage between it, departmental, and individual plans should occur. Targets throughout the organization, driven by customers' needs internally as well as externally, linked from line to staff all striving for quality improvement may sound like Utopia, but in reality it's only Phase V of "implementing quality . . . with a customer focus."

With such attainment, you might assume nothing further remains to be done, but this is not the case. Phase VI is that point in time where quality isn't thought of as something outside the norm—as something extra. It is the accepted way all employees conduct themselves in their daily work. You reach this zenith by following the steps outlined in this book, combining heart and mind, desire and discipline. Keep your mission ever before you and the entire organization, and accept the fact that your pursuit began with education and learning, and so, too, your advancement requires education and learning.

There is one final aspect, one last "to do." It is the providing products and service *beyond* what is needed, *beyond* expectation. You can't justify it today for it seems too costly or impossible. But tomorrow it can and will be done. If not by your organization, by another.

Quality References

More than the customary bibliography usually found in a book's last pages, the following listing includes references I have found particularly useful. You will find these to be mostly of philosophy and theory, but I encourage you to begin your quality pursuit with a good exposure to those who have shaped so much of quality thought. As noted early in this book, you must first develop a feeling about quality that causes you to think about satisfying customer needs in a commonsense way. From these authors you can grasp more than mere fundamentals, and from them you can establish the foundation for building a quality environment in your organization.

Books and Training Materials

Albrecht, Karl, and Ron Zemke. *Service America!* Homewood, IL: Dow-Jones-Irwin, 1985. A book devoted to improvement in the service sector.

Amsden, Davida M., Howard E. Butler, and Robert T. Amsden. *SPC Simplified Workbook: Practical Steps to Quality.* White Plains, NY: Quality Resources, 1990.

Amsden, Robert T., Howard E. Butler, and Davida M. Amsden. *SPC Simplified: Practical Steps to Quality.* White Plains, NY: Quality Resources, 1986.

"Bead Box Exercise." Statco Products, Clawson, Michigan.

Burr, John T. *SPC Tools for Operators.* Milwaukee, WI: ASQC Quality Press, 1989.

Carlzon, Jan. *Moments of Truth.* Cambridge, MA: Ballinger, 1987. Jan Carlzon's bold theory of management and how he reversed the fortunes of three companies.

Deming, W. Edwards. *Out of the Crisis.* Cambridge, MA: Massachusetts Institute of Technology. Center for Advanced Engineering Study, 1982. A text on which the Deming Video Series is based.

Juran, Joseph M. *Managerial Breakthrough: A New Concept of the Manager's Job.* New York: McGraw-Hill, 1964. A fact-filled book which enables managers to perform better when creating or preventing change—the basic function of a manager's job.

Masaaki, Imai. *Kaizen (Ky'zen): The Key to Japan's Competitive Success.* New York: Random House, 1986. The original book about continuous improvement in Japan.

The Memory Jogger. Lawrence, MA: Growth Opportunity Alliance of Greater Lawrence.

Peters, Tom. *Thriving on Chaos: Handbook for a Management Revolution.* New York: Alfred A. Knopf, 1987.

Peters, Tom, and Nancy Austin. *A Passion for Excellence.* New York, Random House, 1985.

Peters, Thomas J., and Robert H. Waterman, Jr. *In Search of Excellence: Lessons from America's Best-Run Companies.* New York: Harper & Row, 1982.

Pfeiffer, J. Williams, and John E. Jones, editors. *A Handbook of Structured Experiences for Human Relations Training*, volume V. San Diego, CA: University Associates, 1975.

Pyzdek, Thomas. *An SPC Primer*. Milwaukee, WI: ASQC Quality Press, 1984.

Pyzdek, Thomas. *Pyzdek's Guide to SPC, Volume I: Fundamentals*. Milwaukee, WI: ASQC Quality Press, 1989.

Resource Engineering, Inc. *SPC in Action: Basic Training for Operators* (training system). White Plains, NY: Quality Resources, 1990.

Walton, Mary. *The Deming Management Method*. New York: Putnam Publishing Group, 1986. W. Edwards Deming and his 14 points of Obligations of Management.

White, B. Joseph. "Accelerating Quality Improvement." Paper presented at the Total Quality Performance Conference, January 21, 1988, sponsored by the Conference Board. Copyright 1988 University of Michigan.

Zemke, Ron, and Dick Schaaf. *The Service Edge: 101 Companies that Profit from Customer Care*. New York: Penguin, Inc., 1989.

Videos

Barker, Joel. "Discovering the Future: The Business of Paradigms." Distributed by Charthouse Learning Corporation.

Blanchard, Ken. "The One-Minute Manager." CBS/Fox Video.

"Dealing with Difficult People from Phoenix." BFA Films & Videos Incorporated.

Houseman, John. "Brainpower." Distributed by Simon & Schuster from Learning Corporation of America.

"I Know It When I See It." AMA Films.

"In Search of Excellence." Distributed by Video Arts, Inc.

Peters, Tom. "A Passion for Customers." Distributed by Video Publishing House, Inc.

"Roadmap for Change—The Deming Approach." Encyclopedia Britannica Education Corporation.

"Team Building." McGraw-Hill Training Systems.

"Tom Peters: The Leadership Alliance." Video Publishing House.

"Working Together Works." Dartnell.

Appendix*

GETTING YOUR TEAM STARTED

(Sample Outline for Team Leaders)

This outline is designed to provide team leaders with specific guidelines for starting team activities. It describes the mechanics of the first few team meetings as well as the responsibilities of the team leader. (An example of an agenda is included for reference.)

The order of events at team meetings will depend upon the type of team involved. Some items will be more important to some teams than they will be to others. However, we suggest that you address all the guidelines. This will help you keep the team focused on your objective.

The workbook from the Leader/Facilitator workshop you participated in describes what is involved in team activity. It contains much useful information. We suggest using it as another reference. Descriptions of the roles of team facilitator, leader, recorder, and advisor have been taken from this Leader/Facilitator workbook and are attached for your reference. You may want to distribute these descriptions to the appropriate team members to help them clarify their responsibilities.

*Developed from materials originally drafted by Ms. Beth McCartney and Ms. Pam Butcher, Citizens Gas & Coke Utility Quality Managers

A. Make sure team supplies and meeting room are available.

1. Team supplies include: masking tape, storyboard cards, watercolor markers, push pins, flipcharts, cork boards, and tripods.
2. A person has been designated at each location to provide you with materials. Please contact the appropriate person.
3. The team recorder should help you to obtain necessary meeting supplies and to schedule the meeting room.

B. Prepare a meeting agenda.

1. Always prepare an agenda prior to each meeting. Distribute it to team members early enough for them to consider meeting topics and complete any "homework" ahead of the meeting. (A sample agenda is included in Attachment A.) The agenda should pose questions for consideration and assign work to be done prior to the meeting. Be sure to indicate the meeting location on the agenda.
2. Stick to the agenda during the meeting.
3. Put unfinished business on the next agenda.
4. Members may suggest agenda topics to the leader.
5. Submit your first meeting agenda to the quality coordinator for his or her review and suggestions. If necessary, meet with the quality coordinator and your facilitator to discuss the agenda.

C. Meet with your recorder and facilitator to discuss their roles (See Attachment B for role descriptions).

D. Review the team's charter/objectives.

1. A charter defining the project scope and the mission statement should be given to team members by the person forming the team.

2. If only a project scope is provided, prepare to develop a team mission by storyboarding or discussing it at a meeting.

3. It is important for your team members to understand why they are on the team and what they are being asked to do. Ask the person chartering the team (the advisor) for help if you do not fully understand your team charter.

E. Conduct a skills inventory.
1. Areas affected by the team objective should be represented on the team. If necessary, provide a part-time consultant for opinions and advice.

F. Establish a team name.
1. The team name should relate to the team's objective or composition; consider using a clever acronym.

G. Establish ground rules.
This is important because it eliminates confusion and problems. The team should discuss and agree on policies for:
1. Attendance and punctuality
   a. Should attendance and punctuality be on the "honor system"?
   b. Should attendance be mandatory?
   c. What should be done about excessively absent team members? Should they be approached by the team leader? Be replaced?
2. Meeting location
   a. Where should meetings be held?
3. Meeting length and frequency
   a. How long should meetings be?
   b. How often should meetings be held?

4. Meeting etiquette
   a. How will team members be treated? (i.e., "Leave your hat at the door; listen to others and let them finish their thoughts," etc.)
5. Outside communication of team activity
   a. How much information about team activity will be communicated to those outside the team? How can we prevent preliminary conclusions from being misunderstood by others?
6. Discussion boundaries
   a. What topics will be discussed at meetings? (Only those topics relevant to the team objective/mission statement?)

H. Establish deliverables.
1. Determine which tangible deliverables the team will develop. Examples include objectives/strategies, recommendations, reports, analyses, etc.
2. Prioritize issues/tasks to aid the discussion and decision-making process.
   a. Be sure to analyze the time and resources required versus the benefits to be gained.

I. Use quality process concepts.
1. Regardless of the team's objective, the customer-satisfaction process should be used. Consider these questions:
   a. What are the team's outputs?
   b. Who is/are our customer(s)?
   c. What are their requirements? (Don't forget to ask them!)
   d. What output should we provide to the customer? (If the team objective is specifically to assess customer needs, these are the exact questions to ask.)

2. Use the six steps to satisfying customers
   a. As problems or issues are encountered, use the process steps to help the team accomplish its objective. In particular, don't let the team get into the solution steps (3, 4, and 5) without first doing the analysis step (2).
3. Decision-making tools such as flowcharts, storyboards, cause-and-effect diagrams, Pareto analyses, histograms, pie charts, cost benefit analyses, surveys, etc., are very helpful for analyzing a problem or issue.

J. Critique each meeting to obtain multiple views and suggestions for improving future meetings.
1. Discuss at the end of the meeting with team members.
2. Meet with the facilitator after team meetings for his or her guidance and feedback.
   a. Share recommendations with the team.
   b. Incorporate ideas.

K. Keep minutes of team meetings.
1. The team recorder is responsible for preparing a summary of each meeting.
   a. To help prepare the minutes, the recorder should use a flipchart, storyboarding cards, or take notes.
2. If the designated recorder is absent, assign minute-taking responsibility to another team member.
3. The team leader is responsible for seeing that minutes are kept and distributed promptly.
4. Minutes should indicate which members are present, what decisions are made, action items, responsibilities, and completion dates for action items.

L. Prepare team status reports.

1. The leader should provide an update of the team's progress to the quality coordinator on a quarterly basis.

M. Implement team ideas.

1. A team does not replace existing organizational/ supervisory authority or responsibilities.

2. Team recommendations should always complement existing decision-making responsibilities.

3. Team recommendations should always be submitted for approval before being implemented. Recommendations can be submitted using memos, reports, or presentations.

4. Utilize the normal procedures requiring managerial approval(s). If issues arise that require special attention, the team leader is responsible for issuing a memo or report to appropriate management stating the team's recommendations.

5. Prepare a final team report upon completion of the team objective/mission.

   a. The final report should summarize the team's efforts and support your deliverables.

   b. Be sure to list specific, tangible results (savings, etc.).

   c. Forward a copy of the final report to the quality coordinator.

6. Management presentations

   a. Have a chairperson or leader.

   b. Introduce everyone.

   c. Have an agenda listing topics and speakers.

   d. Use visual aids—be professional.

   e. Summarize the project.

   f. Emphasize achievements, progress, and accomplishments.

   g. Give costs and benefits.

Attachment A—Sample Agenda

Memorandum

From: D. N. Griffiths   To: J. Calhoun   4/9/91
                        J. Chenoweth
                        J. Clancy
                        W. Diener
                        D. Kaiser
                        F. Lekse
                        C. Lykins
                        W. Ramey
                   cc: D. L. Lindemann
                        Joe Gufreda
                        Beth McCartney
RE: Agenda
     Load Factor Quality Team
     Monday, 4/20/91, 1:30 p.m. to 3:00 p.m.
     Location: First Floor Training Room

The agenda for our first meeting will be as follows:

I. *Introduction:* (DNG)
    A. Review of DLL's memo
       1. Deliverables: Objectives & Strategies
    B. General Mechanics of Team
       1. Roles of facilitator, leader, recorder, members
       2. Meeting length and frequency
       3. Attendance/punctuality
       4. Participation
       5. Skills audit (should others be added to team?)
       6. Use of process

II. *Assessing Customer Needs*
  A. Who are the customers of the team?
  B. What are customer requirements?
  C. What output should we provide?
III. *Mission Statement*
  A. What are the team objectives?
     1. From DLL memo:
        a) Successful resolution of the issue
        b) Demonstrate our commitment
        c) Enhance our individual capabilities
     2. Other?
  B. Does "improve load factor" accurately state our mission?
IV. *Conclusion (no later than 3:00 p.m.)*
  A. Summarize results of today's meeting.
  B. Establish agenda and time of next meeting.
Please come prepared to discuss and storyboard (as appropriate) the agenda items stated.

Attachment B—Roles and Positions of Key Team Personnel

FACILITATOR'S ROLE

—Helps the leader and team adhere to the meeting agenda, time frame, and quality process

—Helps the team accomplish its task(s) by asking questions that maximize analysis

—Provides feedback and reinforcement to the team leader on how effectively the leader and team are using the process and techniques

—Encourages full participation

TEAM FACILITATOR
POSITION DESCRIPTION

Purpose
To help the team accomplish its task by assuring that the team adheres to the meeting agenda and uses the quality process and its techniques. To provide constructive criticism and/or suggestions to the team leader.

Responsibilities
1. Must acquire and demonstrate leadership skills in quality process techniques dealing with customer needs assessment, employee involvement, and issue resolution (problem-solving).

2. Uses questions to challenge the group, summarize discussion, encourage participation, keep discussion moving, etc.

3. Assists, guides, and coaches to ensure effective operation of the team.

4. Communicates problems, concerns, or team needs to the team advisor or quality coordinator.

5. Suggests appropriate tools and techniques to help the team identify or analyze issues.

6. Attends meetings convened by the quality coordinator regarding implementation and process issues.

7. Obtains additional information as required (for "time-outs," presentations, recognition, etc.).

8. Assists other facilitators by conveying information or providing feedback.

LEADER'S ROLE

—Helps team fulfill its charter/objectives

—Acquires necessary knowledge and training

—Recruits team members and monitors their effectiveness

—Determines training needs of team members (as related to team activity)

—Schedules and conducts meetings

—Helps members focus on the process

—Sees that records are kept

—Encourages team involvement

TEAM LEADER
POSITION DESCRIPTION

Purpose
To help the team fulfill its charter/objectives and ensure the smooth and effective operation of the team.

Responsibilities
1. Acquires and demonstrates leadership skills in quality process techniques, especially those dealing with customer-needs assessment, employee involvement, and issue resolution (problem-solving).
2. Schedules, arranges, and conducts meetings. Prepares and follows meeting agendas, schedules meeting location, and makes sure needed materials are available.
3. Seeks required background data (homework) for team members so that all team members are working from the same database.
4. Keeps meetings focused and gets the team "back on track" if necessary.
5. Encourages group involvement by soliciting opinions on various topics from team members.
6. Clearly defines team responsibilities and assignments at the end of each meeting.
7. Makes sure the recorder properly prepares and distributes team minutes.
8. Presents (or arranges to present) team suggestions to management.
9. Makes sure the team's approved ideas are implemented.
10. Coordinates with recorder to keep the quality process coordinator informed of team progress (i.e., team formation/resolution notifications, membership changes, recommendations to management, implementation of team ideas, etc.).

RECORDER'S ROLE

—Keeps a running summary of the meeting by using a flipchart, by storyboarding, or by taking notes.

—Prepares and distributes meeting minutes that summarize:

- Decisions made
- Action items and their completion dates
- Team/member responsibilities

TEAM RECORDER
POSITION DESCRIPTION

Purpose
To keep a running summary of meetings by using a flipchart, by storyboarding, or by taking notes.

Responsibilities
1. To provide minutes of team meetings that summarize decisions made, action items, team/member responsibilities, and completion dates for action items.
2. Works with team leader to make sure minutes accurately reflect team activity and to "set the stage" for future agendas.
3. Distributes minutes to team members.
4. Helps team leader obtain materials and meeting room.
5. Keeps quality process coordinator informed of team activity (i.e., team formation/resolution notifications, membership changes, recommendations to management, implementation of team ideas, etc.).

ADVISOR'S ROLE

—Establishes team charter, mission, objectives, and time benchmarks

—Advises the team on company policies and procedures

—Guides/counsels team as it progresses

—Communicates information to team members

—Counsels, informs, suggests, and recommends (but does not *dictate*)

—Receives team recommendations and *responds to them in a timely fashion*

Index